Fracture Mechanics
Test Methods
for Concrete

Fracture Mechanics Test Methods for Concrete

Report of Technical Committee 89-FMT
Fracture Mechanics of Concrete: Test Methods

RILEM
(The International Union of Testing and Research Laboratories
for Materials and Structures)

Edited by

S.P. Shah
and
A. Carpinteri

CRC Press
Taylor & Francis Group
Boca Raton London New York

CRC Press is an imprint of the
Taylor & Francis Group, an **informa** business
A TAYLOR & FRANCIS BOOK

CRC Press
Taylor & Francis Group
6000 Broken Sound Parkway NW, Suite 300
Boca Raton, FL 33487-2742

First issued in paperback 2019

© 1991 RILEM
CRC Press is an imprint of Taylor & Francis Group, an Informa business

No claim to original U.S. Government works

ISBN-13: 978-0-412-41100-7 (hbk)
ISBN-13: 978-0-367-86589-4 (pbk)
ISBN-13: 978-0-442-31383-8 (USA)

British Library Cataloguing in Publication Data
Available

Library of Congress Cataloging-in-Publication Data
Available

Visit the Taylor & Francis Web site at
http://www.taylorandfrancis.com

and the CRC Press Web site at
http://www.crcpress.com

Contents

RILEM Technical Committee 89-FMT
Fracture Mechanics of Concrete: Test Methods

LIST OF MEMBERS

P. Acker, Laboratoire Central des Ponts et Chaussées, Paris, Cedex 15, France.

Dr M.G. Alexander, Dept. of Civil Engineering, University of the Witwatersrand, Johannesburg, South Africa.

Dr B. Barr, Dept of Civil & Structural Engineering, University of Wales, College of Cardiff, Cardiff, UK.

Dr A. Bascoul, Laboratoire de Genie Civil, INSA-UPS, Toulouse, France.

Professor Z.P. Bazant, Dept. of Civil Engineering, Nothwestern University, Evanston, IL, USA.

Dr Y. Berthaud, Laboratoire de Mècanique et Technologie, ENS de Cachan, Cachan, France.

Dr J.M. Berthelot, Université du Maine, Laboratoire de Mecanique, Le Mans, France.

Mr Brameshuber, Inst. für Massivbau und Baustofftechnologie, Universitat Karlsruhe, Karlsruhe, Germany.

Dr E. Brühwiler, Dept of Civil Engineering, University of Colorado, Boulder, CO, USA.

Professor A. Carpinteri, Dip. Ingegneria Strutturale, Politecnico di Torino, Torino, Italy.

Professor L. Cedolin, Dip. Ingegneria Strutturale, Politecnico di Milano, Milano, Italy.

S. Chhuy, Laborataire Régional de l'Est Parisien, Melun, France.

Professor K.P. Chong, Dept. of Civil Engineering, University of Wyoming, Laramie, Wyoming, USA.

Dr L.S. Costin, Sandia National Laboratory, Albuquerque, NM, USA.

Dr C.H. Detriché, Laboratoire de Genie Civil, INSA-UPS, Toulouse, France.

Professor L. Elfgren, Division of Structural Engineering, Luleå University of Technology, Luleå, Sweden.

Professor M. Elices, Dept de Ciencia de Materiales, ETS Ingenieros de Caminos, Ciudad Universitaria, Madrid, Spain.

Professor R. Eligehausen, Inst. für Werkstoffe im Bauwesen, Universität Stuttgart, Stuttgart, Germany.

Dr G. Ferrara, ENEL, Centro Ricerche Idrauliche Strutturali, Milano, Italy.

M. Fickelson, RILEM, Cachan, France.

V.S. Gopalaratnam, Dept. of Civil Engineering, Columbia, Missouri, USA.

Dr M. Hassanzadeh, Division of Building Materials, Lund Institute of Technology, Lund, Sweden.

Professor N.M. Hawkins, Dept. of Civil Engineering, University of Washington, Seattle, WA, USA.

Professor A. Hillerborg, Division of Building Materials, Lund Institute of Technology, Lund, Sweden.

Professor H.K. Hilsdorf, Inst. für Massivbau und Baustofftechnologie, Universitat Karlsruhe, Karlsruhe, Germany.

D.A. Hordijk, Dept. of Civil Engineering, Delft University of Technology, Delft, The Netherlands.

Dr A. Ingraffea, School of Civil and Environmental Engineering, Hollister Hall, Cornell University, Ithaca, NY, USA.

Professor B.L. Karihaloo, School of Civil & Mining Engineering, University of Sydney, NSW, Australia.

Professor J. Kasperkiewicz, Inst. of Fundamental Technological Research, Academy of Sciences, Warszawa, Poland.

Professor W. Koyanagi, Dept. of Civil Engineering, Gifu University, Yanagido, Gifu, Japan.

Professor V.C. Li, Dept. of Civil Engineering, University of Michigan, Ann Arbor, Michigan, USA.

Doz. Dr H. Linsbauer, Technical University of Vienna (TU-Wien), Vienna, Austria.

Professor J. Mazars, Laboratoire de Mécanique et Technologie, ENS de Cachan, Cachan, France.

Professor H. Mihashi, Dept. of Architecture, Faculty of Engineering, Tohoku University, Sendai, Japan.

Professor S. Mindess, Dept. of Civil Engineering, University of British Columbia, Vancouver, BC, Canada.

Professor B.H. Oh, Dept. of Civil Engineering, Seoul National University, Kwanak-Ku Seoul, Korea.

Dr J. Planas, Dept. de Ciencia de Materiales, ETS Ingenieros de Caminos, Ciudad Universitaria, Madrid, Spain.

Professor H.W. Reinhardt, Institut für Werkstaffe im Bauwesen, Universitat Stuttgart, Stuttgart, Germany.

Professor K. Rokugo, Dept. of Civil Engineering, Gifu University, Yanagido, Gifu, Japan.

Dr P. Rossi, Laboratoire Central des Ponts et Chaussées, Paris, Cedex 15, France.

Dr J. Rots, TNO Institute for Buildings, Materials and Structures, Rijswijk, Delft, The Netherlands.

Mr G. Sawade, Inst. für Werkstoffe im Bauwesen, Universitat Stuttgart, Stuttgart, Germany.

Professor S.P. Shah, Dept. of Civil Engineering, Northwestern University, Evanston, IL, USA.

Professor S.E. Swartz, Dept. of Civil Engineering, Kansas State University, Manhattan, KS, USA.

Professor H. Takahashi, Research Inst. for Strength and Fracture Materials, Faculty of Engineering, Tohoku University, Sendai, Japan.

Dr G. Tognon, Italcementi S.p.A, Bergamo, Italy.

Professor J.G.M. van Mier, Stevinlaboratory, Delft University, Delft, The Netherlands.

Dr M. Wecharatana, Dept. of Civil Engineering, New Jersey Inst. of Techn., Newark, NJ, USA.

Professor F.H. Wittman, Institute for Building Materials, ETH-Hönggerberg, Zurich, Switzerland.

RILEM Secretariat, Pavillon du Crous, 61 av. du Pdt Wilson, 94235 Cachan, France.

RILEM (Réunion Internationale des Laboratoires d'Essais et de Recherches sur les Matériaux et les Constructions) is the International Union of Testing and Research Laboratories for Materials and Structures.

Subcommittees

A. **'Notched beam test: mode I fracture toughness'**
Chairman: Karihaloo
Members: Bascoul, Brameshuber, Carpinteri, Detriche, Elices, Ferrara, Hassanzadeh, Ingraffea, Malvar, Planas, Shah, Swartz, Tognon, Wecharatana.

B. **'Compact specimen testing'**
Chairman: Rossi
Members: Barr, Bascoul, Brameshuber, Brühwiler, Chhuy, Detriche, Karihaloo, Linsbauer, Planas, Rossi, Shah.

C. **'Mixed mode crack propagation'**
Chairman: Carpinteri
Members: Barr, Bazant, Chong, Elices, Ferrara, Hassanzadeh, Hordijk, Ingraffea, Li, Nobile, Swartz, Sawade, Shah, Van Mier.

D. **'Loading rate effects'**
Chairman: Reinhardt
Members: Acker, Bazant, Brühwiler, Mihashi, Mindess, Planas, Rossi, Shah, Wecharatana.

E. **'Fracture process zone detection'**
Chairman: Mindess
Members: Berthaud, Berthelot, Cedolin, Chhuy, Ferrara, Mazars, Rastogy, Shah.

Preface

The purpose of the RILEM Technical Committee 89-FMT 'Fracture Mechanics of Concrete: Test Methods' is to discuss and propose experimental methods to measure the fracture toughness of concrete. Heterogeneous cement-based materials are considered quasi-brittle materials and the conventional test methods based on the classical linear elastic fracture mechanics cannot be applied to the laboratory-sized specimens. A method was proposed by the preceding RILEM Technical Committee 50-FMC. This method was described in *Materials and Structures*, 1985, Vol. 18, pp. 287–290. This method was based on the Fictitious Crack Model which requires three parameters to describe fracture characteristics of concrete: a measure of fracture energy (G_f), tensile strength, and the relationship between crack closing pressure and crack opening displacement. Test recommendations were provided only for determination of G_f. It is now generally recognized that G_f is dependent on the size and geometry of specimens.

Two additional methods were proposed to Committee 89-FMT based on the two-parameter model and the size-effect law. An extensive analysis of these two test methods is reported in Chapter 1. In addition, an effective crack model which is very similar to the two-parameter model is also examined in this chapter. Based on the Committee deliberation, the two test methods: one based on the two-parameter model and the other based on the size-effect law are proposed as recommended procedures. These recommendations appeared in the November 1990 issue of *Materials and Structures*.

Notched-beam specimens are the most common type of specimens used for evaluating fracture parameters. However, other geometries such as compact-tension specimens, are also used. These other methods of determining Mode I fracture parameters are examined in Chapter 2. The wedge loaded cube specimens and cylindrical core specimens described in this chapter are suitable for quality control at the building site and for evaluation of existing structures. These specimens were analyzed using a two-parameter fracture model and using the fictitious crack model.

Crack propagation under mixed-mode loading is discussed in Chapter 3. An extensive review of various test methods is presented. The Committee is currently conducting a round-robin test using a four-point shear specimen. A dimensional analysis of the problem of size effect is presented on the basis of the brittleness number concept. The analysis of crack propagation under compression is discussed in this chapter. Measurements of crack propagation and crack profile using laser holography as well as other studies reported in this chapter point out that friction can play an important role and we need to examine it carefully.

The effects of loading rate on properties such as fracture toughness, fracture energy and crack propagation velocity are discussed in Chapter 4. When designing instruments for applying high rates of loading one must consider inertial effects, stress wave propagation, and local damage. Such considerations and various testing apparatuses are detailed in this chapter. A short review of the fracture theories including the extension of the two-parameter fracture model to predict the rate effect is provided in Chapter 4.

In Chapters 5 and 6, a number of different experimental techniques are described for the observation of the fracture process zone in concrete. It is concluded that acoustic emission and pulse velocity measurements give reasonable results, but that the best results are obtained with laser holographic interferometry techniques. Chapter 6 is specifically devoted to laser interferometry methods.

All the members of the RILEM Technical Committee 89-FMT are gratefully acknowledged for their precious scientific support and for their tireless contribution.

The concepts of fracture mechanics are now making their entry in the international design codes for reinforced and prestressed concrete. The experimental determination of relevant fracture parameters is no longer the subject only of academic research but is also a critical design need. We hope that this book will help the international community in implementing the fracture mechanics-based design codes.

Surendra P. Shah
Chairman,
RILEM Committee 89-FMT,
Northwestern University,
Evanston, Illinois, USA

Alberto Carpinteri
Secretary,
RILEM Committee 89-FMT,
Politecnico di Torino,
Torino, Italy

1 NOTCHED BEAM TEST: MODE I FRACTURE TOUGHNESS[1]

B.L. KARIHALOO and P. NALLATHAMBI
University of Sydney, Sydney, Australia

1 Introduction

This Chapter summarizes the results of an analysis of three-point bend test data according to the proposals based on *the two-parameter model* (Shah and Jenq) and on *the size-effect law* (Bažant). The primary aim of this analysis is to assess the applicability of these two proposals to the determination of the fracture toughness of plain concrete from notched beam specimens.

Of the extensive body of raw test data available for analysis, only a very small proportion was suitable for analysis according to these two proposals. The reasons for this were twofold. First, very few sets of data were accompanied by load-CMOD diagrams necessary for analysis according to the two-parameter model. This may suggest less than universal availability of a servo-controlled testing system. Secondly, even fewer sets of data met the rather strict size requirements of the size-effect law. In order to fully utilize the valuable test data they have also been analysed according to a third proposal, based on the *effective crack model* (Karihaloo and Nallathambi). All available data were amenable to analysis by this proposal.

The contents of this Chapter are structured as follows. A summary of the three proposals and the major recommendation on the determination of Mode I fracture toughness from notched beam tests is given in Section 2. This is followed by self-explanatory copies of the two original proposals (*two-parameter model* in Section 3 and *size-effect law* in Section 4). Tables 1 and 2 summarizing the results of the analysis of test data according to these two proposals appear in the respective sections with suitable notes to make them self-explanatory. Details of the specimen groups summarized in Table 1 are given in Tables 15-21 in Appendix I at the end of the Chapter.

A self-explanatory copy of proposal 3 (the *effective crack model*) forms Section 5. Tables 4 and 5 give a summary of the analysis of test data according to this proposal and a comparison with the results of the previous two proposals. Details of the specimen groups included in the analysis and comparison are given in Tables 22-43 in Appendix II at the end of the Chapter. Section 6 describes the results of an investigation conducted for the sole purpose of comparing the results from the same test specimens according to the *two-parameter* and *effective crack* models

---

[1]With a contribution from M. Elices and J. Planas

using concrete mixes covering a wide range of compressive strengths. Section 7, written in the main by Professors Elices and Planas, develops further the topic of comparing the above proposals on the basis of their asymptotic behaviour in the limit of large size structures. Actually, they limited themselves to a comparison of the two-parameter model and the size effect law with Hillerborg's cohesive crack model using quasi-exponential and linear softening laws. Their methodology was applied by the present authors to the effective crack model as well, to complete the comparison of all three proposals included in this Chapter. This method of comparison addresses to a large extent the comments on the Draft Report from Professor Hillerborg's group.

The Chapter ends with a list of references and two Appendices which give details of the test data assembled from various sources and used in the analysis.

2 Summary and Recommendation

The report describes three independent proposals for estimating the mode I fracture properties of plain concrete from three-point notched bend test data.

The first proposal that is based on the two-parameter model (TPM) requires the registration of load vs crack mouth opening diagram (P-CMOD), including an unloading/reloading sequence at 95% of the peak load (P_{max}) on the descending branch of P-CMOD plot. The initial compliance C_i is used to determine the elastic modulus E of the mix, whereas the unloading compliance C_u is used to estimate the effective Griffith crack length $\underline{a}$ in the test specimen. The critical stress intensity factor, K_{Ic}^s is then calculated from P_{max} and $\underline{a}$. A critical opening displacement of the original pre-crack tip ($CTOD_c$) is also calculated from P_{max} and $\underline{a}$.

The second proposal that is based on the size-effect law (SEL) requires the measurement of just the maximum loads P_j for several notched beams varying only in depth (W_j), but otherwise having the same width, and the same span to depth and notch to depth ratios. The mean fracture energy G_f required for fracture propagation in an infinitely large test specimen is then calculated either numerically or graphically using linear elastic fracture mechanics formulae and a size effect law. The accuracy of the calculated G_f is judged by several statistical measures. An equivalent critical stress intensity factor, designated K_{Ic}^b, may also be calculated from the Griffith relation $K_{Ic}^b = \sqrt{EG_f}$ for comparison with the results of the other two proposals.

The third proposal that is based on the effective crack model (ECM) requires the measurement of just the maximum load P_{max} and the corresponding load point deflection δ_p, preferably on specimens of varying depth W and notch to depth ratio. A set of loads, P_i and corresponding load point deflections δ_i, is also recorded early in the test when the material is still linear elastic. The set of values, P_i, δ_i is used to determine the elastic modulus of the mix E, whereas the peak values, P_{max}, δ_p are used to estimate the effective Griffith crack length, $\dot{a}_e$. The critical stress intensity factor, K_{Ic}^e is calculated using P_{max} and a_e. A more accurate critical stress intensity factor, denoted $\bar{K}_{Ic}^e$ which takes into account the true stress state ahead of a notch in three-point bending, may also be calculated.

The critical stress intensity factors determined according to the three proposals, namely K_{Ic}^s, K_{Ic}^b, K_{Ic}^e (and/or $\bar{K}_{Ic}^e$) have been assigned different superscripts to distinguish them not only one from the other, but also from the commonly used notation K_{Ic} for an ideally brittle material. Roughly speaking, K_{Ic} would represent the microscopic fracture toughness of a quasi-brittle material such as concrete, whereas K_{Ic}^s, K_{Ic}^b, or K_{Ic}^e would represent its macroscopic (and therefore measurable) fracture toughness.

The analysis reported below will show that when tests on notched three-point bend beams are performed in accordance with the requirements of these proposals, all three values of macroscopic fracture toughness are very nearly the same for laboratory-scale specimens. Therefore, it is recommended that any one of them may be used to determine the fracture toughness of plain concrete from three-point bend notched specimens.

The choice of the appropriate method will be dictated by the available testing and specimen preparation facilities. Thus, if it is possible to cast, cure and transport fairly large specimens, the proposal based on *size-effect law* may be followed. If a testing machine is available which can be controlled using a clip gauge output, then the proposal based on the *two-parameter model* would be quite appropriate. In all circumstances, the proposal based on *effective crack model* would be applicable. Whenever possible, fracture toughness should be calculated by more than one method for comparative purposes.

Finally, it is worth remembering that, although the fracture parameters determined according to the three proposals are very nearly the same for laboratory-scale specimens, the load carrying capacities of large structures predicted using these parameters may differ depending on the model used. An attempt has been made in the companion contribution by Elices and Planas (Section 7) to give a reason for this discrepancy.

3 Proposal 1: Two Parameter Model (TPM)

In this proposed recommendation, a direct method is suggested to calculate two size independent fracture parameters, i.e., critical stress intensity factor (K_{Ic}^s) and critical crack tip opening displacement ($CTOD_c$) from the experimental results of single edge notched beams subjected to three point bending under quasi-static loading condition. Crack mouth opening displacement ($CMOD$) has to be monitored in this test to account for the stable crack growth that may occur prior to the peak load. Inelastic displacement due to nonlinear effects should be subtracted so as to apply the LEFM formulae.

3.1 Test specimens

3.1.1 Dimensions

The following specimen dimensions are suggested for maximum aggregate size not larger than 25.4mm (1 inch) (see Fig. 1).

Width(B) × depth(W) × length (L) = 76.2mm × 152.4mm × 711.2mm (3in × 6in × 28in)

Loading span, $S = 609.6$mm (24in)

Notch-depth ratio, $a_o/W = 1/3$

Loading-span/depth ratio, $S/W = 4$

For maximum aggregate size larger than 25.4mm (1 inch), the specimen dimensions should be increased proportionally. At least four specimens are recommended for each type material tested. Sawcut notch or precast notch with notch width less than 3.175mm (1/8in) may be used for the proposed test.

3.1.2 Fabrication of specimens

After casting, the specimens should be covered with wet burlap or kept in the curing room with 100% relative humidity and $23° \pm 2°C$ temperature for the first 24 hours. On the second day all the specimens should be transferred to the curing room until about 4 hours before testing.

3.2 Requirements of test setup

3.2.1 Testing machines

A closed-loop testing machine with $CMOD$ as feedback signal or a relatively stiff machine is recommended to achieve a stable failure. The crack mouth opening displacement ($CMOD$) and the applied load should be recorded continuously during the tests. A clip gauge is recommended to measure the $CMOD$. However, if a clip gauge is not available, LVDT may be used as a replacement. To avoid the error caused by the rotational effect while using LVDT, the gauge length should be kept as small as possible. The $CMOD$ measuring plane should be at the centre of the beam width so as to minimize possible torsional effect.

3.2.2 Rate of test

Since it is a quasi-static test, the rate of the test should be controlled so that the peak load will be reached in about 5 minutes.

3.2.3 Loading and unloading technique

When the specimen is monotonically loaded to post-peak status, it is unloaded when the load is at about 95% of the peak load (Fig. 2). This is done so as to determine the unloading compliance soon after the critical (peak) load.

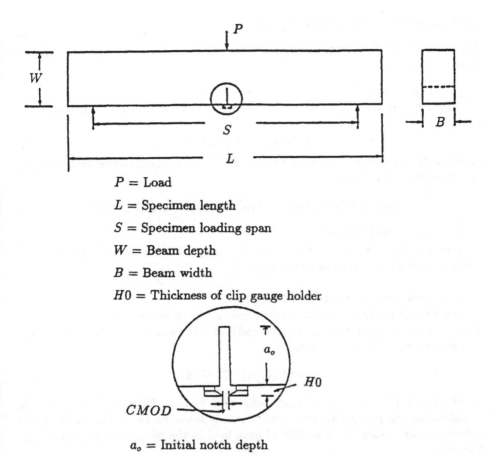

P = Load

L = Specimen length

S = Specimen loading span

W = Beam depth

B = Beam width

$H0$ = Thickness of clip gauge holder

a_o = Initial notch depth

$CMOD$ = Crack mouth opening displacement

Figure 1: Testing configuration and geometry of specimen for TPM.

3.3 Calculation of material properties

A typical load-CMOD plot and composition of $CMOD$ due to nonlinear effect is shown in Figs. 2 and 3, respectively.

3.3.1 Young modulus (E)

The Young modulus (E) is calculated from the measured initial compliance (C_i) using Eqn. 1

$$E = 6Sa_oV_1(\alpha)/(C_iW^2B) \tag{1}$$

in which $S, a, H0, W$, and B are defined in Fig. 1, C_i is equal to the initial compliance (see Figs. 2 and 3),

$$V_1(\alpha) = 0.76 - 2.28\alpha + 3.87\alpha^2 - 2.04\alpha^3 + 0.66/(1 - \alpha)^2 \tag{2}$$

and $\alpha = (a_o + H0)/(W + H0)$.

Based on the finite element analysis, the contribution of the clip gauge holder thickness ($H0$) is included in the function V_1.

3.3.2 Critical stress intensity factor (K_{Ic}^s)

To determine the critical stress intensity factor, the effective critical crack length, $\underline{a}(\underline{a} = a_o+$ stable crack growth at peak load), should be calculated first. The value of $\underline{a}$ is determined by solving Eqn. 3

$$E = 6S\underline{a}V_1(\alpha)/(C_uW^2B) \tag{3}$$

in which $\underline{a}$ = effective critical crack length, $\alpha = (\underline{a} + H0)/(W + H0)$ and C_u = unloading compliance at peak load which is assumed to be the same as the unloading compliance at about 95% of peak load in the post-peak stage (see Figs. 2 and 3).

For those laboratories which cannot perform a stable three-point bend test, the C_u values can be approximately calculated by assuming $CMOD^* = 0$ (Fig. 3). The values of K_{Ic}^s and $CTOD_c$ determined based on this assumption are about 10% to 25% more than the values calculated using the actual unloading compliance.

The critical stress intensity factor is then calculated using Eqn 4

$$K_{Ic}^s = \frac{3P_{max}S}{2BW^2}\sqrt{\pi\underline{a}}F(\alpha) \tag{4}$$

in which $\underline{a}$ = effective critical crack length

$$F(\alpha) = \frac{1}{\sqrt{\pi}}\frac{1.99 - \alpha(1 - \alpha)(2.15 - 3.93\alpha + 2.7\alpha^2)}{(1 + 2\alpha)(1 - \alpha)^{3/2}} \tag{5}$$

$\alpha = \underline{a}/W$, and P_{max} = measured peak load.

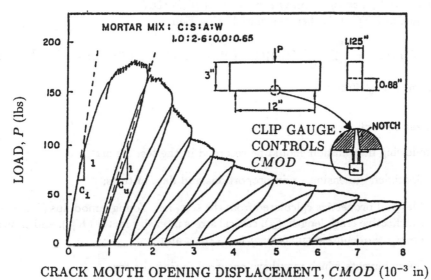

CRACK MOUTH OPENING DISPLACEMENT, $CMOD$ (10^{-3} in)

Figure 2: Typical load - CMOD plot (Jenq & Shah, 1985).

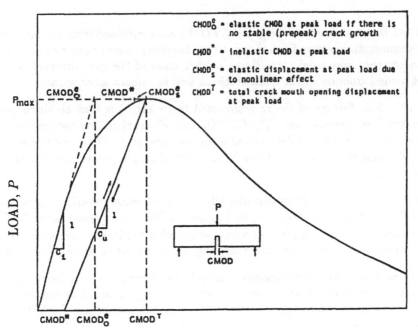

Figure 3: Decomposition of CMOD due to nonlinear effect.

7

3.3.3 Critical crack tip opening displacement (CTOD_c)

The value of $CTOD_c$ is then calculated using Eqn. 6

$$CTOD_c = \frac{6P_{max}Sa}{W^2BE}V_1(\alpha)\{(1-\beta)^2 + (-1.149\alpha + 1.081)(\beta - \beta^2)\}^{1/2} \qquad (6)$$

in which $\alpha = a/W$ and $\beta = a_o/a$.

3.4 Test report

The following information should be recorded in the test report.

1. Specimen dimensions, mix proportions and date of tests;

2. Material properties calculated from the test, i.e., Young's modulus (E), critical effective crack length, (a), critical stress intensity factor (K^s_{Ic}), and critical crack tip opening displacement ($CTOD_c$); and

3. Special events which may have occurred during the test.

3.5 Features of proposed recommendation

The following are some of the distinguishing features of the proposed recommendation.

1. The fracture parameters (K^s_{Ic} and $CTOD_c$) as determined from the proposed recommendation have been shown to be size-independent parameters. These parameters were used correctly to predict most of the experimental results of mode I fracture under quasi-static as well as impact loading rates.

2. One salient feature of the recommended test method is that all three basic material properties (i.e. K^s_{Ic}, $CTOD_c$ and E) used in the two parameter fracture model can be determined from the same test. Thus, the theoretical predictions based on these three basic material properties will be much more consistent.

3. The proposed model is computationally efficient because the critical point and peak load point are clearly defined in the model. Furthermore, since LEFM solutions for some simple geometries are readily available, analysis of these problems can be performed by most engineers even using hand calculators.

4. The proposed fracture parameters have been demonstrated to be useful for applications such as mixed-mode fracture, fibre reinforced concrete, reinforced concrete beams subjected to shear, strain rate effects, and high strength concrete.

3.6 Applications

Various applications of the two fracture parameters K_{Ic}^s and $CTOD_c$ to some practical structural problems and strain-rate effects are demonstrated by Jenq and Shah (1985), John, Shah and Jenq (1987), and John and Shah (1987).

3.7 Analysis of test data

Table 1 summarizes the results of available test data according to the two-parameter model. Details of the specimen groups summarized in this table are given in Tables 15-21 in Appendix I at the end of the Chapter.

Table 1. Summary of results according to proposal 1 (Shah and Jenq)

Serial No.	g (mm)	No. of specimens tested	$K_{Ic}^s(MPa\sqrt{m})$		$CTOD_c(mm)$		For details see table in Appendix I
			Mean	sd	Mean	sd	
1	19	6	0.931	0.263	0.0148	0.0076	15
2	19	4	1.054	-	0.0153	-	15
3	19	3	1.128	0.269	0.0200	0.0085	15
4	19	25	1.146	0.014	0.0316	0.0067	16
5	19	22	1.220	0.102	0.0312	0.0087	17
6	3	12	0.894	0.068	0.0042	0.0013	18
7	6	2	1.141	0.095	0.0145	0.0057	19
8	13	3	1.475	0.191	0.0220	0.0086	19
9	13	3	1.530	0.022	0.0169	0.0024	19
10	19	8	0.976	0.103	0.0170	0.0068	20
11	32	17	1.211	0.121	-	-	21
12	2	11	0.790	0.090	-	-	21
13	8	2	2.130	-	0.0338	0.0039	20

The entries in Table 1 have been grouped according to mix variables only because they do not vary with the size of test specimens. Thus entries differ only by the maximum size of coarse aggregate (g) used in the mix and other mix parameters, e.g. water/cement ratio, texture of coarse aggregate. (Mix properties are given in the respective detailed tables 15 - 21 in Appendix I.)

Fig 4 shows the variation of relative K_{Ic}^s with specimen depth. The relative K_{Ic}^s of a mix is calculated by dividing the K_{Ic}^s of a particular specimen group from this mix by the K_{Ic}^s for the specimen group of least depth from the same mix. Thus, the K_{Ic}^s values of specimens with depth 305 mm and 203 mm in Table 15 (there is very little difference in mix properties) have been divided by average K_{Ic}^s value of the specimen group with depth 102 mm. On the other hand, if specimens of only one depth have been tested, then no relative K_{Ic}^s can be calculated. This explains the absence of entries from Tables 16, 17, and 19. The numbers in parentheses denote

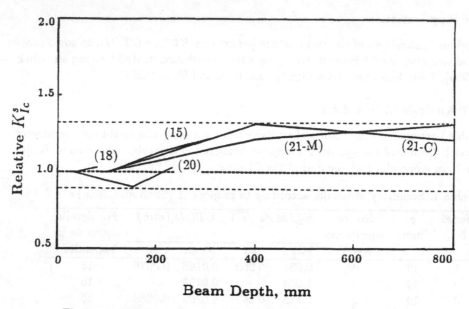

Beam Depth, mm

Figure 4. Variation of relative K^s_{Ic} values with specimen depth.

the corresponding table numbers from Appendix I and the hyphenated numbers denote the particular mix groups within the Table.

3.8 Conclusions

1. The fracture toughness according to two-parameter model K^s_{Ic} is essentially independent of specimen depth. It varies between 0.93 and $1.53 MPa\sqrt{m}$ for normal concrete and between 0.79 and $0.89 MPa\sqrt{m}$ for mortar. This variation is mainly due to the variation in the mix properties.

2. The two-parameter model requires the solution of a fourth-order non-linear equation for the calculation of $\underline{a}$. This can greatly reduce its practical usage. John, Shah and Jenq (1987) have recently provided nomograms for estimating $\underline{a}$ which overcomes this drawback.

3. In order to test specimens according to the two-parameter model a sophisticated closed-loop servo-controlled machine is required.

4 Proposal 2: Size Effect Law (SEL)

4.1 Scope and definition

The recommendation covers the size effect method proposed by Bažant (1987) for determining the fracture energy of concrete by measuring the maximum loads of geometrically similar notched concrete specimens of different sizes. The fracture energy G_f which is obtained by this method is defined (Bažant, 1987) as the specific energy (i.e. energy per unit crack plane area) required for fracture growth in an infinitely large test specimen. The fracture energy defined in this manner is not the same (and cannot be the same) as that obtained by other methods. According to Bažant (1987) and Bažant and Pfeiffer (1986, 1987), this definition is, in theory, independent of both the specimen size and shape, and the value obtained gives correct results when used in finite element codes.

The method is described here for Mode I fracture only, although the same approach is possible for Mode II and Mode III shear fractures.

4.2 Specimens

4.2.1 Material
The specimens of all sizes must be cast from the same batch of concrete. The quality of concrete must be as uniform as possible.

The entire curing procedure and the environments to which the specimens are exposed, including their histories, must be the same for all the specimens until the time of the test.

4.2.2 Shape of specimens
Although the present method works equally well for specimens of different shapes (Bažant and Pfeiffer 1986, 1987), three-point bend beams are recommended for the purpose of standardization. The recommended shape of specimen is shown in Fig. 5. It is a beam of width B, depth W, and length L. The beam is loaded at midspan by a concentrated load, and is simply supported over span S. A notch of depth a_o is cut into the cured beam by a diamond saw at midspan (Fig. 5). The loads are applied though one hinge and two rollers with the minimum possible rolling friction. The concentrated load as well as the support reactions, which represent uniformly distributed line loads across the beam width, are applied over bearing plates whose thickness d_1 is such that they could be considered as rigid. The bearing plates are either glued with epoxy or are set in wet cement. The distance from the end of the beam to the nearest support must be sufficient to prevent spalling and cracking at beam ends.

The span-to-depth ratio of the specimen, S/W, should be at least 2.5. The ratio of the notch depth to the beam depth, a_o/W, should be between 0.15 and 0.4. The notch width should be as small as possible and should not exceed 0.5 of the maximum aggregate size, g. The width, B, and the depth, W, must not be less than $3g$.

Specimens of at least three different sizes, characterized by beam depths $W = W_1, \ldots W_n$, and spans $S = S_1, \ldots S_n$ must be tested. The smallest depth W_1 must not be larger than $5g$, and the largest depth W_n must not be smaller than $15g$. The ratio of maximum W to minimum W must be at least 4. The ratios of the adjacent sizes should be approximately constant. Optimally, the size range should be as broad as feasible. Thus, e.g., the choice $W/g = 4, 8, 16$ is usually acceptable, but the choice $W/g = 3, 6, 12, 24$ is preferable.

At least three identical specimens should be tested for each specimen size.

Ideally, the specimens of all sizes should be geometrically similar in two dimensions, with the third dimension (width B) the same for all specimens[2]. This means that the ratios $S/W, a_o/W$, and L/W should be the same for all specimens.

It is nevertheless possible, although at the cost of introducing an additional experimental error, to use specimens for which only the cross sections are geometrically similar in the dimensions (i.e., a_o/W is the same) while the spans (and lengths) are dissimilar.

4.3 Apparatus

It is sufficient to use an ordinary uniaxial testing machine without servo-control and without high stiffness. However, servo-control and a high stiffness of the loading frame lead to more consistent results. The machine must be capable accurately to register the maximum load. The post-peak response need not be measured, but if it is, one obtains further useful data which are valuable, e.g., for calibration of a finite element model or for comparison with other methods.

4.4 Test procedure

The specimens should be loaded at constant or almost constant displacement rate. As a normal type of test, the loading rates should be such that the maximum load is reached within 1 to 10 minutes, however, other loading rates may be used for special purposes. The displacement rate is optimally chosen so that the theoretical velocity of the crack tip at maximum load is the same for all the specimen sizes. However, this requires sophisticated calculations prior to testing, and in their absence one may choose displacement rates such that the maximum load is achieved for all specimen sizes within approximately the same time; this is approximately achieved when the load-point displacement rates are proportional to $R = S^2 W^{-1.25}$ (explanation: this is a compromise between linear elastic fracture mechanics, for which $R = S^2 W^{-1.5}$, and the strength criterion, for which $R = S^2 W^{-1}$).

---

[2]Variations of thickness may affect the fracture energy value due to three-dimensional nature of the elastic singularities at specimen surface as well as the shear lip phenomenon. They may also introduce undesirable size effects of thickness due to hydration heat as well as drying.

4.5 Test results

The test results that are needed to determine the fracture energy are only the maximum load values $P_1, \ldots P_n$ for specimens of various sizes $W_1, \ldots W_n$. The following data should also be reported: all dimensions of the beams and of the bearing plates; maximum aggregate size; the ratios (by weight) of water, cement, sand, and gravel in the mix; type of cement, its fineness, admixtures, mineralogical type of aggregate, curing conditions, storing conditions, temperature and humidity during the test, standard cylinder strength of concrete and its Young's modulus, and the mean mass density of concrete for each specimen.

4.6 Calculation procedure and example

1. If L_j is almost the same as S_j,

$$P_j^0 = P_j + \frac{1}{2} m_j g \qquad\qquad (j = 1, \ldots n) \qquad (7)$$

   in which m_j is the mass of specimen j, g = gravitational acceleration, and n = number of tests.

   If L_j is much larger than S_j,

$$P_j^0 = P_j + \frac{2S_j - L_j}{2S_j} m_j g \qquad\qquad (j = 1, \ldots n) \qquad (8)$$

2. If geometrically similar specimens are used, set $P_j^* = P_j^0$ and go to step 3. If not, then for each beam depth W_j calculate the value (Eqn. 8 of Bažant, Kim and Pfeiffer, 1986):

$$P_j^* = P_j^0 \frac{W_m S_j}{W_j S_m} \qquad\qquad (j = 1, \ldots n) \qquad (9)$$

   where W_m/S_m is the depth-span ratio for the mid-size specimen. Preferably though, avoid structures where P_j^* differs from P_j^0 by more than 15%.

3. Now carry out linear regression, considering the plot of the ordinates Y_j versus the abscissae $X_j = W_j$ where

$$Y_j = \left(\frac{BW_j}{P_j^*} \right)^2, \qquad (10)$$

   Determine the slope A of the regression line:

$$A = \frac{\sum_j (X_j - \bar{X})(Y_j - \bar{Y})}{\sum_j (X_j - \bar{X})^2} \qquad (11)$$

13

where

$$\bar{X} = \frac{1}{n} \sum_j X_j, \qquad \bar{Y} = \frac{1}{n} \sum_j Y_j \qquad (12)$$

$(\bar{X}, \bar{Y})$ is the centroid of all data points.

Also check whether the plot of data points is approximately linear. If not, then some errors or disturbing effects have probably occurred in the testing procedure.

4. Calculate auxiliary values for the extrapolation to very large specimen sizes for which linear elastic fracture mechanics applies. According to handbooks by Tada et al. (1985) or Murakami et al. (1987):

for $S/W = 4$ (with $\alpha_0 = a_o/W$):

$$F_4(\alpha_0) = 1.090 - 1.735\alpha_0 + 8.20\alpha_0^2 - 14.18\alpha_0^3 + 14.57\alpha_0^4 \qquad (13)$$

for $S/W = 8$:

$$F_8(\alpha_0) = 1.107 - 1.552\alpha_0 + 7.71\alpha_0^2 - 13.55\alpha_0^3 + 14.25\alpha_0^4 \qquad (14)$$

Then, by linear interpolation,

$$F(\alpha_0) = F_4(\alpha_0) + \frac{(S_m/W_m) - 4}{4} [F_8(\alpha_0) - F_4(\alpha_0)] \qquad (15)$$

Now, evaluate the nondimensional energy release rate:

$$g(\alpha_0) = \left(\frac{S_m}{W_m}\right)^2 \pi\alpha_0[1.5F(\alpha_0)]^2 \qquad (16)$$

When $S/W < 3$ or $S/W > 10$, the interpolation is not sufficiently accurate and $g(\alpha_0)$ must be calculated by linear elastic fracture mechanics. ($F(\alpha_0)$ can be determined for arbitrary specimen shape by a linear elastic finite element code, even one without singular finite elements.)

5. Now, calculate the fracture energy (mean prediction):

$$G_f = \frac{g(\alpha_0)}{E_c A} \qquad (17)$$

Remark 1. - It is also possible to estimate the characteristic size of the fracture process zone using the Y-intercept of the regression plot (Bažant et al., 1986).

6. Finally, calculate the statistics:

$$s_X^2 = \frac{1}{n-1} \sum_j (X_i - \bar{X})^2, \qquad s_Y^2 = \frac{1}{n-1} \sum_j (Y_j - \bar{Y})^2 \qquad (18)$$

$$s_{Y|X}^2 = \frac{1}{n-2} \sum_j (Y_j - Y_j')^2 = \frac{n-1}{n-2}(s_Y^2 - A^2 s_X^2) \qquad (19)$$

$$\omega_{Y|X} = s_{Y|X}/\bar{Y}, \qquad\qquad \omega_X = s_X/\bar{X} \qquad (20)$$

$$\omega_A = \frac{s_{Y|X}}{As_X\sqrt{n-1}} \qquad\qquad m = \frac{\omega_{Y|X}}{\omega_X} \qquad (21)$$

and the approximation:

$$\omega_G^2 = \omega_A^2 + \omega_E^2 \qquad (22)$$

where $j = 1, 2, ...n$; X_j, Y_j are the measured data points (not averages of some groups of data points); $Y_j - Y_j'$ are the vertical deviations of these points from the regression line, ω_X = coefficient of variation of the sizes chosen, $\omega_{Y|X}$ = coefficient of variation of the errors (vertical deviations from the regression line), ω_A = coefficient of variation of the slope of the regression line, m = relative width of scatterband, ω_E = coefficient of variation of the values of the elastic modulus E_c used in Eqn. 17. If the values of E and G_f are assumed to be uncorrelated, the coefficient of variation of G_f is $\omega_{G_f} \approx \omega_G$, and if they are assumed to be perfectly correlated, $\omega_{G_f} \approx \omega_A$. In reality, $\omega_A < \omega_{G_f} < \omega_G$. If the values of E_c have not been measured they may be estimated from the measured strength values f_c' by using some established empirical formula.

The value of ω_A, should not exceed about 0.08 and the value of m about 0.15. These conditions prevent situations in which the size range used is insufficient compared to the scatter of results. Such a situation is illustrated in Fig. 8(a), in which a unique regression slope is obtained but the slope A is highly uncertain. Fig. 8(b) illustrates the case where a large scatter of test results necessitates the use of a very broad size range, while Fig. 8(c) illustrates the case where a small scatter of test results permits the use of a not very broad size range. Since the value of ω_A can be made small by testing very many specimens (high n) even if the scatterband is wide, it is necessary also to limit the value of m, in addition to limiting the value of ω_A. Obviously, the necessary size range can be reduced by carefully controlled testing which results in a low scatter.

4.6.1 Example (from Bažant and Pfeiffer 1986, 1987)

For a concrete of maximum aggregate size 0.5 in., the following data have been measured using three-point bend specimens (Fig. 5) of width $B = 1.5$ in. $S/W = 2.5$, $L/W = 8/3$ and notch depth $a_o = W/6$. $E_c = 4,010,000 psi$, $g(\alpha_0) = 6.37$.

Measured Maximum Loads (Fig. 6):

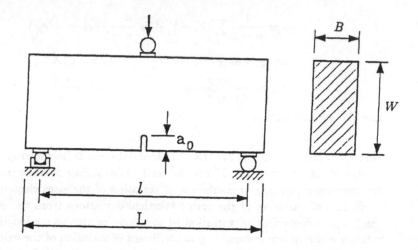

Figure 5: Testing configuration and geometry of specimen for SEL.

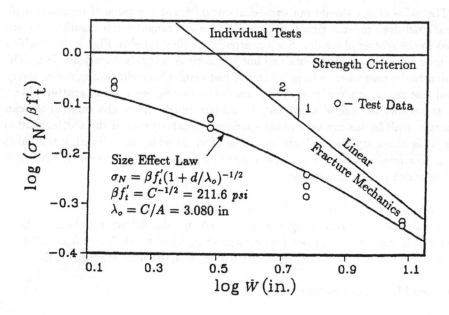

Figure 6: Size effect plot constructed from the max. load values measured.

| Depth W (in.) | Max. Loads (lb.) | | | Mean P (lb.) |
	1	2	3	
1.5	405	408	417	410
3.0	677	706	711	698
6.0	990	1040	1096	1042
12.0	1738	1739	1773	1750

Data for regression (Fig. 7):

$X = W$ (in.) $Y = (BW/P)^2$

1.5	3.09,	3.04,	2.91	x	10^{-5}
3.0	4.42,	4.06,	4.01	x	10^{-5}
6.0	8.26,	7.49,	6.74	x	10^{-5}
12.0	10.73,	10.71,	10.31	x	10^{-5}

Regression line $Y = AX + C$ (Fig. 7):

$$
\begin{aligned}
A &= 7.253 \text{ x } 10^{-6} \, in.^{-1} psi^{-2} \\
C &= 22.340 \text{ x } 10^{-6} \, psi^{-2} \\
s_{Y|X} &= 6.882 \text{ x } 10^{-6} \, in.^{-1} psi^{-2} \\
s_X &= 4.644 \, in. \\
\omega_{Y|X} &= 0.1090, \\
\omega_X &= 0.8526,
\end{aligned}
$$

Result:

$$
G_f = \frac{6.37}{(7.253 \times 10^{-6})(4.01 \times 10^6)} = 0.219 \, lb./in.
$$

$$
\omega_A = 0.0616, \qquad m = 1.1315
$$

Note: 1 in. = 25.4mm, 1 lb. = 4.448 N, 1 psi = 6895 Pa

4.7 Caveat: Other size effects

Scattered results can, however, also arise when other size effects are present. This is of particular concern for relatively large and thick specimens, in which hydration heat can cause different heating in small and large specimens and thus lead to a size effect that is superimposed on that described by the size effect law. In that case the present method fails. Similarly, it can fail when size effects of different types are produced by drying, which affects small specimens differently from the large ones.

4.8 Background

For materials in which the fracture front is blunted by a nonlinear zone of distributed cracking or damage, the fracture energy G_f may be uniquely defined

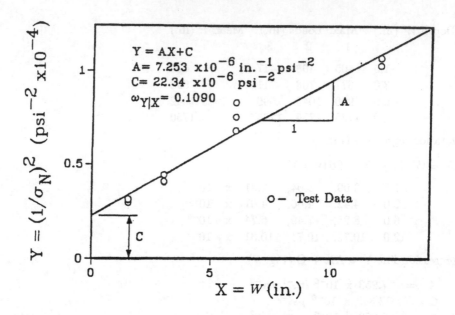

Figure 7: Linear regression plot constructed from the max. load values measured.

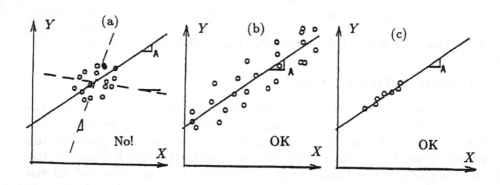

Figure 8: Examples to construct regression plot.

(Bažant and Pfeiffer, 1986, 1987) as the energy required for crack growth in an infinitely large specimen. This definition is, of course, independent of specimen size. It has been proven theoretically that this definition must lead to the same fracture energy values for various possible geometrical shapes of specimens. This definition avoids the chief problem with other methods of determining fracture energy, which are known to be more or less sensitive to the specimen size as well as geometrical shape.

The foregoing definition is practically useful only if the law for extrapolating to infinite specimen size is known. Although the exact form of the size effect law for blunt fracture is not known, an approximate form which appears to be sufficient for practical purposes is given by Bažant's size effect law (Bažant, 1984):

$$\sigma_N = \beta f_t' \left(1 + \frac{W}{\lambda_o g}\right)^{-1/2} \tag{23}$$

in which σ_N = nominal stess at failure, and β, λ_o = empirical constants.

Evaluation of experimental results according to Eqn. 17 has been shown to yield G_f -values which are approximately the same for very different specimen geometries, including edge-notched tension specimens, three-point bend notched specimens, and eccentric compression notched specimens. Taking into account the inevitable scatter of test results for a material such as concrete, it appears that the size effect law in Eqn. 23 is adequate for a size range such as approximately 1:30, which is sufficient for most practical purposes.

For a still broader size range, a more complicated form of the size effect law is required, as shown recently by Elices and Planas (private communication, 1986) who calculated the size effect curves very accurately for a line crack with a postulated stress displacement relation ahead of the crack tip. From such analyses, as well as other finite element studies with the crack band model, it appears that for a broader size range the size effect law in Eqn. 23 becomes insufficient. A more complex form of the size effect law can be formulated; however, the additional terms appear to depend on the specimen shape as well as other material properties.

An advantage of the method is its simplicity. It is sufficient to measure only the maximum loads of the test specimens, which means that even a relatively soft testing machine or a machine without displacement control can be used. Compared to certain other methods, it is not necessary to measure the location of the crack front, which is a notoriously difficult problem. Neither is it required to measure the crack opening displacements. Measurement of the unloading stiffness is also unnecessary.

The fracture energy values obtained by the present method have been shown by Bažant and Pfeiffer (1986, 1987), to be consistent with the fracture energy values obtained by finite element fitting of test results.

4.9 Analysis of test data

A summary of the results of analysis according to Proposal 2 is given in Table 2. As previously mentioned, just these few data sets were amenable to analysis according to the size-effect law.

Table 2. Summary of results according to Proposal 2 (Bažant and Pfeiffer)

Sl. No.	$g^\S$	¶	Size, mm S B W	a_0/W	$A^\dagger$	G_f (J/m)	$K^{b\,*}_{Ic}$	$\omega_A^\star$	$m^\star$	Ref.
1		6	0600 80 076							
2		4	1000 80 140							
3	20	4	1200 80 200	0.200	0.759E-01	22.986	0.874	0.144	0.161	
4		4	1500 80 240							
5		4	1800 80 300							
6		6	0600 80 076							
7		4	1000 80 140							
8	20	4	1200 80 200	0.300	0.123E-00	23.989	0.892	0.185	0.211	A
9		4	1500 80 240							
10		4	1800 80 300							
11		6	0600 80 076							
12		4	1000 80 140							
13	20	4	1200 80 200	0.400	0.231E-00	21.600	0.847	0.102	0.121	
14		4	1500 80 240							
15		4	1800 80 300							
16		3	095 38 038							
17		3	191 38 076							
18	13	3	381 38 152	0.167	0.578E-02	42.531	1.084	0.133	0.146	
19		3	762 38 305							B
20		3	095 38 038							
21		3	191 38 076							
22	5	3	381 38 152	0.167	0.873E-02	23.670	0.883	0.031	0.048	
23		3	762 38 305							
24		3	0400 100 100							
25	19	2	0800 100 200	0.400	0.270E-01	71.059	1.520	0.537	0.585	
26		2	2000 100 500							
27		8	0400 100 100							C
28		2	0800 100 200							
29	19	2	1200 100 300	0.200	0.709E-02	94.058	1.751	0.331	0.435	
30		1	2000 100 500							
31		1	3200 100 800							

Notes:

1. Of the hundreds of specimen groups for which peak load values were available only the above few groups satisfied the rather strict size requirements of proposal 2. Even

for these few groups, statistical measures point towards poor quality of results (see note 3 below).

2. † Slope of regression line. Care should be exercised in determining A because G_f and therefore K_{Ic}^b are very sensitive to this parameter. A slight error in the calculation of A can significantly alter G_f.

3. * For the purposes of comparison G_f has been converted to a fracture toughness value $K_{Ic}^b, (MPa\sqrt{m})$ using the LEFM relationship $K_{Ic}^b = \sqrt{G_f E}$.

4. ⋆ According to proposal 2, ω_A should not exceed 0.08 and m should be about 0.15. It is evident that only one group of specimens gives good quality results according to these statistical measures. Even a slight deviation from the size requirements results a significant difference in the measured values of G_f (see, e.g. Sl. No. 24–31).

5. A - Nallathambi (1986), Nallathambi and Karihaloo (1986 a, b); B - Bažant and Pfeiffer (1987); C - Alexander (1987).

6. § Max. aggr. size in mm; ¶ Number of specimens tested.

4.10 Conclusions

1. The fracture toughness according to the *size-effect law*, K_{Ic}^b is easy to evaluate. It varies between 0.85 and $1.75 MPa\sqrt{m}$ for normal concrete. Although there was no overlap between the sets of test data that were amenable to analysis by this as well as the two-parameter method, it would seem that K_{Ic}^b compares favourably with K_{Ic}^s for similar, if not identical, concrete.

2. In order to obtain an accurate estimate of fracture toughness according to the size-effect law it would seem essential that the test specimen dimensions meet all the size requirements listed in the proposal. It may not always be easy to meet these requirements especially if they lead to the testing of very large specimens, e.g. when large size coarse aggregate is in use. Moreover, extreme care should be exercised in determining the slope A of the regression line, as was also emphasized by Professor Swartz in his comments on the Draft Report.

3. The above conclusions should be read in conjunction with the conclusions reached in the contribution by Professors Elices and Planas (Section 7.4).

5 Proposal 3: Effective Crack Model (ECM)

In the present proposal, a direct method is suggested to determine the critical stress intensity factor (K_{Ic}^e) from the experimental results of single edge notched beams tested in three-point bending under quasi-static loading condition. The fracture parameter K_{Ic}^e calculated according to this proposal is essentially independent of the specimen size. LEFM formulae are used after the initial notch depth has been suitably augmented to reflect the non-linear load deflection behaviour prior to the attainment of the peak load. Apart from the simpler method of determination of the size of the effective crack, this proposal is similar in principle to proposal 1 (TPM).

5.1 Test specimens

The following specimen dimension ranges are suggested. No critical linear dimension should be less than five times the maximum aggregate size (g) used which may be in the range of 5 – 25 mm. For instance the depth (W) of the specimen for a 20 mm maximum aggregate size mix should be such that the uncracked ligament size is at least 100 mm. The width (B) may be between 40 and 100 mm and the loading span/depth ratio (S/W) may be between 4 and 8. The notch/depth ratio (a_0/W) may be between 0.2 and 0.6 but preferably either 0.3 or 0.4.

At least four specimens are recommended for each type of material tested. Sawcut notch or precast notch (using a well greased wedge) of width not greater than 3 mm and milled to a sharp edge may be used.

After casting, specimens should be covered with wet burlap or kept in a curing room with 100% relative humidity and $23 \pm 2°C$ temperature for the first 24 hours. On the second day all the specimens should be stripped from their moulds (and any wedges used for precast notches gently removed) and transferred to the curing room until about an hour before testing. The specimens should be measured before testing.

5.2 Requirements of test setup

Any relatively stiff machine may be used to load the specimens incrementally while simultaneously measuring the deflection adjacent to the notch. While measuring the deflection at the midspan care should be exercised to avoid any support disturbances. To achieve this, additional fixtures may be necessary to attach the measuring device, such as transducer to the specimen, as shown in Fig. 9. Plot load against midspan deflection readings up to the peak load. Alternatively, if a more sophisticated machine is available allowing a closed-loop set up then the CMOD (measured with a clip gauge) or the loadpoint displacement (measured with a LVDT) may be used as a feedback signal to achieve a stable failure. It is how-

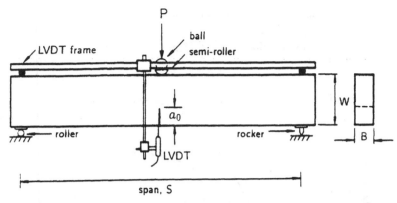

Figure 9: Loading apparatus and fixing arrangement for LVDT.

ever, emphasized that the present proposal does not require any load/deflection information past the peak load. The load should be incremented so that the peak load is attained in 1-10 minutes.

5.3 Calculation of material properties

Determine the Young modulus (E) of the mix by testing the cylindrical specimens in accordance with your Code requirements. It is however preferable to measure strains using two electrical strain gauges with gauge length 50mm glued to diametrically opposite sides at mid-height of a standard cylinder (ϕ150mm). Note that the determination of E using the cylindrical specimen is required only when a continuous record of loads and deflections up to the maximum load cannot be kept.

5.3.1 Effective critical crack length (a_e)
Typical load-deflection plots up to the peak load are shown in Fig. 10, for various a_0/W ratios. Read the values of P_i, P_{max} and the corresponding mid-span deflections δ_i, δ_p (Fig. 10). Calculate E of the specimen from the following equation

$$E = \frac{P_i}{4B\delta_i}\left(\frac{S}{W}\right)^3\left[1 + \frac{5wS}{8P_i} + \left(\frac{W}{S}\right)^2\left\{2.70 + 1.35\frac{wS}{P_i}\right\} - 0.84\left(\frac{W}{S}\right)^3\right]$$
$$+ \frac{9}{2}\frac{P_i}{B\delta_i}\left(1 + \frac{wS}{2P_i}\right)\left(\frac{S}{W}\right)^2 F_2(\alpha_0), \tag{24}$$

where w is the self-weight of the beam, and

$$F_2(\alpha_0) = \int_0^{\alpha_0} \beta\, F_1^2(\beta)\, d\beta, \tag{25}$$

with $\alpha_0 = a_0/W$, and for $S/W = 4$

$$F_1(\beta) = \frac{1.99 - \beta(1 - \beta)(2.15 - 3.93\beta + 2.70\beta^2)}{(1 + 2\beta)(1 - \beta)^{\frac{3}{2}}} \tag{26}$$

23

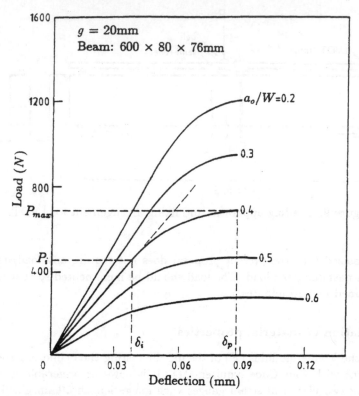

Figure 10: Typical load-deflection plots up to peak load for various a_0/W ratios.

However, a slightly less accurate expression (error $< 1\%$) for $F_1(\beta)$ is available in the range $0.1 < \beta < 0.6$ for two span to depth ratios ($S/W = 4$ and 8) with linear interpolation permitted within, and outside of, these ratios

$$F_1(\beta) = A_0 + A_1\beta + A_2\beta^2 + A_3\beta^3 + A_4\beta^4 \tag{27}$$

where

$$
\begin{aligned}
A_0 &= +0.0075\frac{S}{W} + 1.90 \\[4pt]
A_1 &= +0.0800\frac{S}{W} - 3.39 \\[4pt]
A_2 &= -0.2175\frac{S}{W} + 15.40 \\[4pt]
A_3 &= +0.2825\frac{S}{W} - 26.24 \\[4pt]
A_4 &= -0.1450\frac{S}{W} + 26.38
\end{aligned}
\tag{28}
$$

The coefficients $A_i(i = 0, 1, ..., 4)$ have been obtained from the coefficients given by Brown and Srawley (1966) for $S/W = 4$ and 8 by linear interpolation.

The reduction in the stiffness of the beam (Fig. 10) is a result of both the stable crack growth and the formation of the discontinuous process zone ahead of the visible crack. It is however difficult to separate these two causes. Therefore it is assumed that the critical notch depth a_e may be calculated by introducing a fictitious beam containing a notch a_e whose unchanged stiffness (proportional to E) would be equal to the reduced stiffness of the real beam containing a notch of depth a_0, i.e.

$$\delta_p = \frac{P_{max}}{4BE}\left(\frac{S}{W}\right)^3\left[1 + \frac{5wS}{8P_{max}} + \left(\frac{W}{S}\right)^2\left\{2.70 + 1.35\frac{wS}{P_{max}}\right\} - 0.84\left(\frac{W}{S}\right)^3\right]$$
$$+ \frac{9}{2}\frac{P_{max}}{BE}\left(1 + \frac{wS}{2P_{max}}\right)\left(\frac{S}{W}\right)^2 F_2(\alpha_e), \tag{29}$$

where

$$F_2(\alpha_e) = \int_0^{\alpha_e} \beta F_1^2(\beta)d\beta \tag{30}$$

Here $\alpha_e = a_e/W$ and $F_1(\beta)$ is given by Eqn. 26.

From Eqn. 29 α_e is calculated by a trial and error procedure as follows. For a given notched beam, the initial modulus E is calculated from Eqn. 24. Using $\alpha_0 = a_0/W$ of this beam as a starting value, $F_2(\alpha_e)$ is calculated from Eqn. 30 and substituted into Eqn. 29 together with the corresponding P_{max} and δ_p to calculate a fresh value of E, say E^*. (In the first iteration, E^* will naturally be less than E). The notch depth is incremented by $\Delta\alpha = 0.001$, and the procedure repeated until $E^* = E \pm \varepsilon(= 0.5\%)$. The value of α at this final step is taken equal to α_e.

This procedure was applied to each factorial group of test beams (Appendix II) and the corresponding a_e/W was calculated. The following regression equation gives the best fit

$$\frac{a_e}{W} = \gamma_1\left(\frac{\sigma_n}{E}\right)^{\gamma_2}\left(\frac{a_0}{W}\right)^{\gamma_3}\left(1 + \frac{g}{W}\right)^{\gamma_4}, \tag{31}$$

where g is the maximum size of the aggregate used in the mix, $\sigma_n = 6M/(BW^2)$, $M = (P_{max} + wS/2)S/4$, and $\gamma_1 = 0.088 \pm 0.004, \gamma_2 = -0.208 \pm 0.010, \gamma_3 = 0.451 \pm 0.013$ and $\gamma_4 = 1.653 \pm 0.109$. The elastic modulus E in Eqn. 31 was determined from the initial region of the load/deflection plot using Eqn. 24. This method of determining E is more consistent with the effective crack model.

There is however no reason to suspect that the model is less accurate if the elastic modulus E appearing in Eqn. 29 is determined from separate tests, say on cylindrical specimens. For the values of E measured from cylindrical specimens using electrical strain gauges of sufficient gauge length (usually at least three times the maximum aggregate size, g), the regression Eqn. 31 is still applicable but with $\gamma_1 = 0.198 \pm 0.015, \gamma_2 = -0.131 \pm 0.011, \gamma_3 = 0.394 \pm 0.013$ and $\gamma_4 = 0.600 \pm 0.092$.

5.3.2 Critical stress intensity factor (K_{Ic}^e)
The critical stress intensity factor is then calculated using

$$K_{Ic}^e = \sigma_n\sqrt{a_e}F(\alpha) \tag{32}$$

where $\alpha = (a_e/W)$ and $F(\alpha)$ is given by Eqns. 26 or 27.

5.4 Improved expressions for K_{Ic}^e

It is possible to improve the above expression K_{Ic}^e by considering the true state of stress at the front of a pre-crack in a three- point bend specimen. Elastic finite element calculations show that this state of stress consists not only of a tensile stress normal to the crack faces (as assumed in deriving Eqn. 32) but also of a significant (tensile) stress in the plane of the crack and of a shear stress. By making an allowance for the true stress state ahead of the crack front, Eqn. 32 becomes

$$\bar{K}_{Ic}^e = \sigma_n \sqrt{a_e} Y_1(\alpha) Y_2(\alpha, \beta) \tag{33}$$

where as before $\alpha = a_e/W$, $\beta = S/W$, and

$$Y_1(\alpha) \quad = \quad A_0 + A_1\alpha + A_2\alpha^2 + A_3\alpha^3 + A_4\alpha^4 \tag{34}$$
$$Y_2(\alpha, \beta) \quad = \quad B_0 + B_1\beta + B_2\beta^2 + B_3\beta^3 + B_4\alpha\beta + B_5\alpha\beta^2 \tag{35}$$

Regression coefficients $A_i, B_j, (i = 0, ..., 4; j = 0, 1, ..., 5)$ are given in Table 3.

Table 3. Regression coefficients $A_i, B_j(i = 0, ..., 4; \; j = 0, ..., 5)$

i/j	0	1	2	3	4	5
A_i	3.6460	-6.7890	39.2400	-76.8200	74.3300	-
B_j	0.4607	0.0484	-0.0063	0.0003	-0.0059	0.0033

Kasperkiewicz (1990) performed an analysis of data from five tables chosen at random from the more than twenty tables given in Appendix II and found that

$$\bar{K}_{Ic}^e = 0.004697 + 1.137822 K_{Ic}^e, \tag{36}$$

with a coefficient of correlation $r = 0.998$. For practical purposes, he found that the above relation could be replaced with

$$\bar{K}_{Ic}^e = 1.138 K_{Ic}^e, \tag{37}$$

without any loss of accuracy.

5.5 Test report

The test report should contain the following information:

1. Test specimen dimensions, mix properties, dates of casting and testing and method of introducing the initial notch.

2. Material properties which may have been determined from cylindrical specimens, namely Young's modulus (E) and compressive strength (f_c').

3. Plots of load vs midspan deflection recorded automatically or drawn from recorded readings, indicating P_{max}, P_i and the corresponding δ_p, δ_i.

4. Fracture parameter K_{Ic}^e (or $\bar{K}_{Ic}^e$) calculated from three-point bend specimens.

5. Any special events which may have been observed during testing.

5.6 Distinguishing features of proposed recommendation

1. The fracture parameter K_{Ic}^e (or $\bar{K}_{Ic}^e$) has been shown to be essentially independent of the specimen dimensions.

2. The test is simple to perform using a standard universal testing machine available in most laboratories. No feedback control is required.

3. Calculation of effective notch depth, a_e and the fracture parameter K_{Ic}^e is straightforward.

5.7 Additional information

Additional information on the proposed recommendation may be obtained from Nallathambi and Karihaloo (1986 a, b). (Note: K_c and G_c have been used in these publications in place of K_{Ic}^e and G_{Ic}^e). Note also that the regression formulas for calculating a_e, namely Eqn. 31 for this proposal differ from our published expression (Nallathambi and Karihaloo, 1986 a,b). The formula given in this proposal has two major advantages over our published expressions. First, it has been derived on the basis of not only our test data but also that of many other investigators worldwide. Secondly, it does not depend on the type of the test specimen. It could equally well be used with compact tension data provided the nominal stress is properly interpreted.

5.8 Analysis of test data

The test data of three-point bend specimens from several researchers, together with the authors' own extensive test data were analysed to determine the effective crack depth a_e and hence K_{Ic}^e ($\bar{K}_{Ic}^e$). Whenever possible, the predicted fracture toughness values were compared with K_{Ic}^s and/or K_{Ic}^b according to proposals 1 and 2. A summary of the analysis results is given in Tables 4 and 5, which are provided with self-explanatory notes. Details of the few specimen groups summarized in Table 5 and of the many groups summarized in Table 4 are given in Tables 22-43 in Appendix II at the end of the Chapter.

Fig 11 shows the variation of relative K_{Ic}^e with specimen depth. The relative K_{Ic}^e was calculated by dividing the K_{Ic}^e of a particular specimen group with the K_{Ic}^e of the smallest specimen group from the same mix.

27

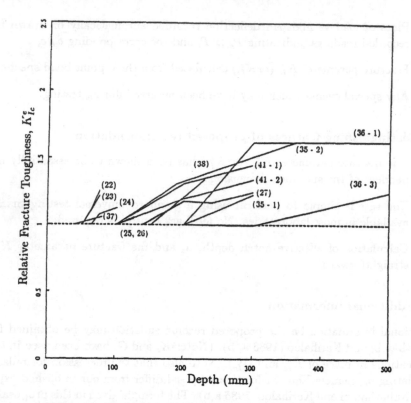

Figure 11: Variation of relative K_{Ic}^e with specimen depth

The numbers in parentheses denote the corresponding table numbers from Appendix II and the hyphenated numbers denote the particular mix groups within the table. The plots from tables 35, 36 and 41 were estimated from the given peak loads. The variation within the same table of different mix groups, especially with tables 35 and 36 suggests poor quality of data. With the remaining plots the relative K_{Ic}^e is reasonably independent of the size of the specimen. It is expected that the variation will further reduce when tests are conducted according to the requirements of the proposal.

5.9 Conclusions

1. The fracture toughness according to effective crack model K_{Ic}^e is essentially independent of beam depth provided the net ligament size exceeds about five times the maximum size of the coarse aggregate (e.g. for a particular coarse aggregate type of maximum size 20mm the net ligament depth should be at least 100mm).

2. K_{Ic}^e is in good agreement with K_{Ic}^s (Table 4) and K_{Ic}^b (Table 5).

3. Calculation of fracture toughness according to the effective crack model would seem to be the easiest of the three proposals, in their present form.

Table 4. Summary of results according to Proposals 1 (TPM) and 3 (ECM)

Sl. No.	g (mm)	No. of specn. tested	K_{Ic}^e $(MPa\sqrt{m})$ Mean	sd	K_{Ic}^e $(MPa\sqrt{m})$ Mean	sd	K_{Ic}^s $(MPa\sqrt{m})$ Mean	sd	For details see table in Appendix II
1	2	90	0.633	0.065	0.727	0.072			22
2	5	84	0.641	0.057	0.738	0.064			23
3	10	60	0.706	0.046	0.780	0.050			24
4	14	30	0.728	0.018	0.818	0.018			25
5	20	30	0.776	0.028	0.881	0.030			26
6	20	76	0.884	0.057	1.022	0.073			27
7	19	6	-	-	-	-	0.931	0.263	30
8	19	3	0.965	0.045	1.099	0.044	1.054	-	30
9	19	3	1.145	0.098	1.330	0.082	1.128	0.269	30
10	19	20	0.797	0.075	0.903	0.082	1.146	0.014	31
11	19	20	0.971	0.078	1.108	0.093	1.220	0.102	32
12	10	12	0.760	0.039	0.885	0.044			33
13	10	8	0.908	0.068	1.045	0.075			33
14	10	8	0.859	0.091	0.993	0.099			33
15	10	15	1.023	0.052	1.167	0.047			34
16	20	14	1.031	0.129	1.200	0.148			35
17	16	15	1.120	0.217	1.286	0.249			35
18	19	8	1.413	0.272	1.600	0.300			36
19	19	4	1.759	0.172	2.004	0.179			36
20	19	16	1.232	0.147	1.397	0.163			36
21	3	12	0.926	0.061	1.045	0.073	0.894	0.068	37
22	6	2	1.221	0.061	1.376	0.069	1.141	0.095	38
23	13	3	1.429	0.046	1.607	0.050	1.475	0.191	38
24	13	3	1.610	0.026	1.811	0.029	1.530	0.022	38
25	13	12	0.975	0.150	1.053	0.163			43
26	5	12	1.004	0.068	1.084	0.075			43
27	19	8	0.982	0.169	1.118	0.192	0.976	0.103	39
28	13	6	0.421	0.091	0.503	0.103			40
29	8	11	1.198	0.199	1.393	0.232			41
30	8	11	0.830	0.117	0.972	0.138			41
31	12	6	1.356	0.101	1.586	0.118			41
32	12	6	0.727	0.038	0.854	0.045			41
33	32	17	1.585	0.210	1.858	0.245	1.211	0.121	42
34	2	11	1.102	0.098	1.280	0.115	0.790	0.090	42
35	8	2	1.896	0.064	2.158	0.071	2.130	-	39

Notes:

1. The entries have been grouped according to mix variables only because they do not vary with the size of test specimens. Thus entries differ only by the maximum size of coarse aggregate (g) used in the mix and other mix parameters, e.g. water/cement

29

ratio, texture of coarse aggregate. (Mix properties are given in the respective detailed tables in Appendix II.)

2. Absence of an entry in K_{Ic}^e column means that load-crack mouth opening plot was not available.

Table 5. Summary of results according to Proposals 2 (SEL) and 3 (ECM)

Gr.⊙ No.	a_0/W	$A^\dagger$	G_f (J/m)	K_{Ic}^b *	ω_A *	m *	K_{Ic}^e Mean(sd)	$\bar{K}_{Ic}^e$ Mean(sd)
G-1	0.200	0.759E-01	22.986	0.874	0.144	0.161		
G-2	0.300	0.123E-00	23.989	0.892	0.185	0.211	0.867(0.063)	1.005(0.076)
G-3	0.400	0.231E-00	21.600	0.847	0.102	0.121		
G-4	0.167	0.578E-02	42.531	1.084	0.133	0.146	0.975(0.150)	1.053(0.163)
G-5	0.167	0.873E-02	23.670	0.883	0.031	0.048	1.004(0.068)	1.084(0.075)
G-6	0.400	0.270E-01	71.059	1.520	0.537	0.585	1.264(0.293)	1.433(0.327)
G-7	0.200	0.709E-02	94.058	1.751	0.331	0.435	1.208(0.260)	1.416(0.369)

Notes:

1. ⊙ Details of individual specimen size, coarse aggregate size used in the mix, and no. of specimens tested are given in Table 43 in Appendix II.

2. All K_{Ic} values are given in $MPa\sqrt{m}$.

3. Of the hundreds of specimen groups for which peak load values were available only the above few groups satisfied the rather strict size requirements of proposal 2. Even for these few groups, statistical measures point towards poor quality of results (see note 6 below).

4. † Slope of regression line. Care should be exercised in determining A because G_f and therefore $K_{Ic}^b(MPa\sqrt{m})$ are very sensitive to this parameter. A slight error in the calculation of A can significantly alter G_f.

5. * For the purposes of comparison G_f has been converted to a fracture toughness value K_{Ic}^b using the LEFM relationship $K_{Ic}^b = \sqrt{G_f E}$.

6. ⋆ According to proposal 2, ω_A should not exceed 0.08 and m should be about 0.15. It is evident that only one group of specimens gives good quality results according to these statistical measures. Even a slight deviation from the size requirements results in a significant difference in the measured values of G_f (see, e.g. Group Nos. 6 and 7).

6. Comparison of ECM and TPM

As seen from Table 4 the fracture parameters of plain concrete calculated using the effective crack model(ECM) and the two parameter model(TPM) are in good agreement despite the fact that they were determined from essentially separate series of test data. This was because of the paucity of simultaneous load-displacement and load-CMOD measurements from one and the same three-point bend specimen. Despite this drawback, Kasperkiewicz (1990) found from an analysis of data from five tables chosen at random from the more than twenty tables given in Appendix II that K_{Ic}^s and K_{Ic}^e are reasonably well correlated (coefficient of correlation $r = 0.832$). To check whether or not this correlation can be improved, an investigation was conducted in which both the load-displacement and load-CMOD plots were simultaneously recorded for all notched specimens tested in three-point bending. A further aim of this investigation was to make this comparison for a wide variety of concrete mixes, ranging in (cylinder) compressive strength from about $25MPa$ to nearly $80MPa$.

It is found that irrespective of the concrete strength, the fracture parameters calculated using the effective crack model (namely, the effective fracture toughness K_{Ic}^e and the effective traction-free notch length a_e) are practically indistinguishable from the corresponding parameters calculated using the two parameter model (namely, the fracture toughness K_{Ic}^s and the notch length a; the latter being also used to determine the critical crack tip opening displacement($CTOD_c$)).

6.1 Specimen preparation and testing

To determine fracture parameters according to ECM and TPM tests were conducted on concrete beam specimens in three-point bending. Five mixes were designed for the target compressive strength, f_c' varying from $20MPa$ to $80MPa$ with coarse aggregate of maximum size 20mm. The mix proportions for all the five mixes and their designations are shown in Table 6. The mix $C5$ required the use of a water reducing agent (superplasticizer) in order to increase its workability. Crushed basalt and ordinary Portland cement (Type A) were used in the mixes.

The specimen size and geometry were selected according to the size requirements of ECM. Specimens of 200mm depth, 80mm breadth and 900mm length were cast in steel moulds. A steel wedge of thickness 6mm, and length 60mm milled to a sharp end was fixed with the mould in order to form a precast notch (to a notch/depth ratio, $a_0/W = 0.3$). From each mix four beam and two cylinder (ϕ150mm) specimens were cast and covered with polyethylene sheets. The specimens were stripped on the following day and placed in the fog room until tested.

All specimens were tested on the twenty-eighth day in a moist state in a displacement controlled closed loop machine (Instron). The load was measured by an electrical load cell and the deflection by a Linearly Varying Displacement Trans-

Table 6. Mix proportions (kg/m^3)

Mix Designation	C1	C2	C3	C4	C5
Basalt, 20mm	1127.2	1108.5	1079.4	1027.9	952.5
Basalt, 5mm	393.2	386.6	376.5	358.5	332.2
Sydney Sand	284.6	279.9	272.5	259.5	240.5
Fly ash	86.7	85.2	83.0	79.0	73.2
Type A Cement	223.1	271.6	347.2	480.9	732.9
Water	204.7	201.3	196.0	186.7	155.2
Superplasticizer, $m\ell$	-	-	-	-	4000.0
Yield	2319.4	2333.2	2354.6	2392.6	2486.4
Aggregate : Cement	6.9	5.8	4.6	3.2	2.0
Water : Cement	0.77	0.64	0.50	0.36	0.20
Slump, mm	100.0	100.0	100.0	75.0	100.0

ducer (LVDT). The $CMOD$ was measured using a clip gauge mounted on a knife edge of thickness 3.5mm and was used as the feedback signal for the displacement control. The speed was selected such that a $CMOD$ of 1mm was achieved in 10 minutes. The output from the load cell, LVDT and clip gauge were connected to an $X-Y$ recorder to obtain the full load-deflection and load-CMOD plots simultaneously on the same chart. After the test the actual specimen dimensions were measured. The geometrical details of the specimens are given in Table 7.

6.2 Test results

6.2.1 Elastic modulus

Table 8 shows the average compressive strength, f_c' of each of the five mixes as obtained from tests on at least two cylinders. The table also shows the elastic modulus E of each of the mixes determined using strain gauges (gauge length = 60mm, i.e. three times the maximum aggregate size). Also shown are the tensile strength f_t estimated from the well-known empirical relation $f_t = 0.4983\sqrt{f_c'}\,(MPa)$ and the elastic modulus, E_c estimated from the empirical relation $E_c = 4734\sqrt{f_c'}\,(MPa)$. It is seen that the measured E compares very favourably with the estimated E_c.

Fig. 12 shows typical load-displacement ($P-\delta$) and load-CMOD plots for one of the five mixes. The $P-\delta$ and $P-CMOD$ plots have been displaced on both scales for clarity. The initial (linear) portions of these plots were used to calculate the elastic modulus, E, and the results are shown in Table 8. It is seen that all four values of E for each of the mixes are in excellent agreement.

6.2.2 Critical load and Compliance measurement

Table 9 summarizes the various experimentally measured quantities required for the calculation of fracture parameters according to ECM and TPM. Included in the table are the peak load, P_{max}, δ_p - deflection at peak load and the unloading compli-

Table 7. Test Data - Specimen Details

Sl. No.	Specimen Designation	Specimen Size, mm $S \times B \times W$	a_0/W
1	C1-1	800×81×203	0.296
2	C1-2	800×81×204	0.294
3	C1-3	800×81×204	0.294
4	C2-1	800×81×203	0.296
5	C2-2	800×81×203	0.296
6	C2-3	800×81×203	0.296
7	C2-4	800×81×203	0.296
8	C3-1	800×80×203	0.296
9	C3-2	800×80×203	0.296
10	C3-3	800×81×204	0.294
11	C4-1	800×82×204	0.289
12	C4-2	800×80×203	0.296
13	C4-3	800×81×204	0.294
14	C4-4	800×81×204	0.294
15	C5-1	800×80×203	0.296
16	C5-2	800×80×203	0.296
17	C5-3	800×82×203	0.289
18	C5-4	800×82×204	0.289

Table 8. Mix Properties

Mix	Compressive Strength $f'_c(MPa)$	Elastic Modulus, (GPa)				Tensile Strength $f_t^{\odot}(MPa)$
		E^*	E_c^{**}	$E^{\dagger}$	$E^{\ddagger}$	
C1	26.8	24.62	24.51	25.56(.35)	25.04(.29)	2.58
C2	39.0	33.80	29.56	29.87(.21)	31.56(.64)	3.11
C3	49.4	34.65	33.27	33.28(.22)	32.96(.24)	3.50
C4	67.5	37.20	38.89	37.13(.23)	38.39(.82)	4.09
C5	78.2	40.30	41.86	40.99(.60)	40.26(.99)	4.41

Notes:
* Determined from separate cylinder tests (using strain gauges)
** Estimated from the relationship, $E_c = 4734\sqrt{f'_c}MPa$ $(=57000\sqrt{f'_c}\text{psi})$
† Calculated from $P - \delta$ plot
‡ Calculated from $P - CMOD$ plot
⊙ Estimated from the relationship, $f_t = 0.4983\sqrt{f'_c}MPa$ $(=6\sqrt{f'_c}\text{psi})$

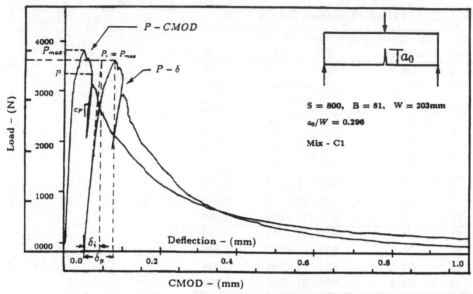

Figure 12: Typical $P - \delta$ and $P - CMOD$ plots for Mix C1.

ance, C_u corresponding to $0.95P_{max}$, together with δ_i, P_i $(= P_{max})$ an arbitrary deflection and corresponding load in the elastic range and C_i - the initial compliance. In order to calculate δ_i, a point on the $P - \delta$ plot at around $0.5P_{max}$ was selected and a straight line was drawn through that point from the origin to P_{max} along the initial straight portion of the $P - \delta$ plot. The projected length on the X-axis was measured as δ_i, and P_i was taken to be equal to P_{max}. This procedure is sketched in Fig. 12. In connection with the measurement of C_u it should be pointed out that a certain degree of inaccuracy is unavoidable here because of the serious practical difficulties experienced in commencing unloading at precisely $0.95P_{max}$ beyond the peak load. To compensate for this approximate load level at the commencement of unloading, the unloading compliance C_u has been corrected by a linear interpolation technique, whereby $C_u = C_i + (C_P - C_i) * 0.05P_{max}/(P_{max} - P)$. Here, P is the actual load at which the unloading was commenced and C_P is the corresponding compliance. The corrected C_u values are given in Table 9.

6.2.3 Calculated fracture parameters[1]
The fracture parameters calculated according to proposals 1 and 3 are listed in

---

[1]The two ECM fracture parameters (Δa_e and K_{Ic}^e), together with a suitable approximation to the transmitted stress-displacement relation in the tension-softening zone have been successfully used (Karihaloo and Nallathambi, 1989 c, Nallathambi and Karihaloo, 1990) to estimate f_{t_1} w_c and process zone length which are required for a full description and finite element implementation of the fracture process.

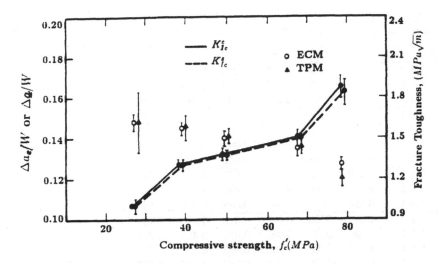

Figure 13: Variation of $\Delta a_e/W$, $\Delta a/W$, K_{Ic}^e and K_{Ic}^s with f_c'

Table 10. The variation of Δa_e, Δa, and K_{Ic}^e and K_{Ic}^s with f_c' is shown in Fig. 13. As mentioned before, the agreement between the two proposed models is excellent. It should however be mentioned that the regression formula (Eqn. 31, proposal 3) overestimates slightly the values of a_e/W (see Table 10). It is therefore recommended that a_e be calculated from Eqn. 29 of proposal 3 as explained earlier.

7 Application of Proposed Fracture Models to Large Structures[2]

This companion contribution (Sections 7.1-7.4) to the main Report will show first that, whilst the fracture parameters based on TPM and SEL are able accurately to predict the maximum load capacity of concrete specimens in the practical range of sizes, the predictions for large structures differ by as much as 31%. It will then argue that the reason for this discrepancy lies in the different definitions of size effect used in these models and in the different fracture energies resulting from the size effect. The methodology used for this comparison is based on the behaviour of the various models (TPM, SEL, as well as Hillerborg's fictitious crack model, FCM with quasi-exponential or linear softening) at the asymptotic limit of large size structures for which the LEFM is strictly valid. In a slight departure from the rest of this Report, the comparison has been made not only for the notched three-point bend geometry but also for the compact tension (CT) and single-end notch (SEN) tension geometries. Results of an experimental investigation designed to apply to TPM, SEL and FCM will also be presented.

[2]Sections 7.1 - 7.4 by M. ELICES and J. PLANAS, The Technical University, Madrid, SPAIN

Table 9. Measured test data

No.	Specimen	a_0/W	$P_{max}^{\dagger}$ (N)	δ_i (mm)	δ_p (mm)	C_i	C_u
						$mm/N \times 10^{-5}$	
1	C1-1	0.296	3840	0.052	0.080	0.600	1.210
2	C1-2	0.294	4120	0.056	0.084	0.584	1.240
3	C1-3	0.294	4000	0.055	0.082	0.601	1.480
4	C2-1	0.296	5200	0.060	0.091	0.467	1.020
5	C2-2	0.296	5150	0.060	0.090	0.471	1.060
6	C2-3	0.296	5100	0.060	0.090	0.475	1.020
7	C2-4	0.296	5120	0.060	0.090	0.493	1.020
8	C3-1	0.296	5450	0.058	0.087	0.464	0.981
9	C3-2	0.296	5660	0.060	0.089	0.464	0.964
10	C3-3	0.294	5900	0.060	0.088	0.446	0.946
11	C4-1	0.289	6720	0.060	0.088	0.378	0.792
12	C4-2	0.296	6000	0.057	0.082	0.406	0.828
13	C4-3	0.294	6500	0.060	0.088	0.375	0.755
14	C4-4	0.294	6380	0.058	0.085	0.382	0.793
15	C5-1	0.296	7500	0.065	0.093	0.392	0.743
16	C5-2	0.296	7780	0.065	0.094	0.378	0.728
17	C5-3	0.289	8600	0.070	0.100	0.354	0.649
18	C5-4	0.289	9170	0.073	0.103	0.343	0.647

$\dagger$ As read from the $P - \delta$ plot

Table 10. Fracture parameters

Specimen	a_0/W	a_e/W	a_e/W (Eqn. 31)	$\underline{a}/W$	K_{Ic}^e	K_{Ic}^s	$CTOD_c$ (mm)
					$MPa\sqrt{m}$		
C1-1	0.296	0.450	0.451	0.429	0.980	0.923	0.0471
C1-2	0.294	0.440	0.444	0.436	1.013	1.001	0.0242
C1-3	0.294	0.439	0.447	0.464	0.982	1.055	0.0284
C2-1	0.296	0.443	0.453	0.444	1.287	1.294	0.0275
C2-2	0.296	0.440	0.454	0.449	1.264	1.298	0.0266
C2-3	0.296	0.440	0.455	0.440	1.252	1.253	0.0258
C2-4	0.296	0.440	0.454	0.433	1.257	1.232	0.0253
C3-1	0.296	0.439	0.450	0.437	1.348	1.343	0.0254
C3-2	0.296	0.436	0.446	0.434	1.386	1.380	0.0263
C3-3	0.294	0.430	0.443	0.437	1.393	1.419	0.0266
C4-1	0.289	0.426	0.436	0.429	1.546	1.559	0.0257
C4-2	0.296	0.426	0.448	0.431	1.426	1.446	0.0240
C4-3	0.294	0.430	0.441	0.427	1.531	1.516	0.0232
C4-4	0.294	0.430	0.443	0.432	1.504	1.514	0.0237
C5-1	0.296	0.423	0.435	0.417	1.758	1.730	0.0267
C5-2	0.296	0.426	0.432	0.420	1.838	1.808	0.0256
C5-3	0.289	0.416	0.422	0.404	1.913	1.851	0.0257
C5-4	0.289	0.412	0.416	0.409	2.015	2.000	0.0266

Sections 7.5-7.6 extend this methodology to the comparison of effective crack (ECM) and fictitious crack models (FCM) for notched three-point bend geometry. It will be shown that the predictions for large structures differ by no more than about 17%.

7.1 Asymptotic behaviour of TPM, SEL and FCM

The comparison procedure for cohesive models is based on the maximum loads predicted by them for geometrically similar specimens of different sizes. The tool is a general expression for the size effect applicable to all cohesive crack models which is derived from their asymptotic properties, as reviewed below.

The application of cohesive crack models to concrete fracture was pioneered by Hillerborg (1976) more than a decade ago (see also Carpinteri, 1980, 1984). The present contribution is based on cohesive crack models with no bulk dissipation. An extensive review of these models can be found in a recent RILEM report (Elices and Planas, 1989).

In all cohesive crack models, fracture under monotonically increasing mode I loading is assumed to occur when the maximum (tensile) principal stress reaches the tensile strength of the material f_t. However, fracture is assumed to be localized in the process zone and is modelled by a displacement discontinuity with the proviso that the faces of the discontinuity are capable of transmitting certain cohesive stresses, less than f_t, such that $\sigma = F(w)$ with $F(0) = f_t$ and $F(w) \geq 0$, where $F(w)$ describes the softening behaviour.

In practice, $F(w)$ is assumed to vanish when the crack opening displacement w reaches a certain critical value w_c. The fracture energy, G_F is then defined as the energy required to open a unit crack fully, or in other words, as the area under the tension-softening curve between $w = 0$ and $w = w_c$. Moreover, from the material parameters E, ν, f_t and G_F two independent parameters are defined; one with the units of length, called the characteristic length $l_{ch} = G_F E/f_t^2$ and the other with the units of displacement, called the characteristic crack opening $w_{ch} = G_F/f_t$.

Now, we note that if the process zone size remains bounded when the size D of a precracked body is allowed to increase without bound, then all cohesive models must converge to linear elastic models. At the same time, however, the observed nominal stress at failure will decrease as D increases, so long as the precracked bodies of increasing D are geometrically similar. Based on dimensional analysis and LEFM (which is rigorously applicable when $D \to \infty$), Planas and Elices (1986a) have demonstrated the following general expression for the maximum load size effect

$$EG_F(K_{INmax})^{-2} = 1 + B_1 l_{ch}/D + B_2(l_{ch}/D)^2 + \ldots \tag{38}$$

with l_{ch} defined above, B_i are dimensionless coefficients depending on the geometrical ratios and shape of $F(w)$, and

$$K_{INmax} = \sigma_{Nmax}\sqrt{D}S(a_0/D) \tag{39}$$

37

where $S(a_0/D)$ is the shape function and a_0 the precrack length in body D.

A subsequent asymptotic analysis performed by Elices and Planas (1986b, 1987, 1990) showed that the asymptotic expression for the size effect (38) may be written as

$$EG_F(K_{INmax})^{-2} = 1 + \left(2\frac{S_0'}{S_0}\right)\Delta a_{C\infty}\,D^{-1} + 0\left(\frac{1}{D}\right) \tag{40}$$

where the critical effective crack extension $\Delta a_{C\infty}$ depends only on $F(w)$, but is independent of f_t and G_F. The dependence of $\Delta a_{C\infty}$ on $F(w)$ is rather strong, especially on the shape of the trailing part of the softening curve. In (40) S_0 refers to $S(a_0/D)$ and S_0' is its first derivative with respect to its argument.

From (40) it follows that two cracked samples of the same material are asymptotically equivalent when the factor $(2S_0'/S_0)D^{-1}$ has the same value for both samples. Consequently, we may define an intrinsic size D_i (Planas and Elices, 1988b) as

$$D_i^{-1} = (2S_0'/S_0)D^{-1} \tag{41}$$

At first glance it may appear that D_i depends on the selection of D. But it is easily shown not to be so, because (41) is equivalent to

$$D_i^{-1} = \frac{\partial \ln K_I^2}{\partial a}\Big|_{a=a_0}$$

where the partial derivative is with respect to crack length a, whereas the load and all remaining dimensions of the specimen remain constant. It should be borne in mind that this analysis was done only for positive geometries (i.e. when the stress intensity factor increases with crack length, or $S_0' > 0$).

When the specimens considered are all geometrically similar (homothetical) we have a single degree of freedom and the selection of D is immaterial. However, when we consider more degrees of freedom the arbitrariness in the selection of the size makes it difficult to compare the results, and the difficulty increases if we consider different specimens and loadings. Therefore, the introduction of an objective measure, the intrinsic size, seems appropriate. With this definition, (38) takes the simpler form:

$$EG_F(K_{INmax})^{-2} = 1 + \Delta a_{C\infty}\,D_i^{-1} + 0\left(\frac{1}{D_i}\right) \tag{42}$$

A general size-effect relation is sketched in Fig. 14 for cohesive cracks, together with the asymptotic expression (42), and the limiting cases applicable for small sizes (limit analysis) and very large sizes (LEFM).

7.1.1 Specific fracture energy

It is useful to define the specific fracture energy relative to the model and experimental procedure used for its determination. This may be accomplished by identifying G_F with $G_{F,mod,exp}$. Moreover, it is essential to indicate that bulk

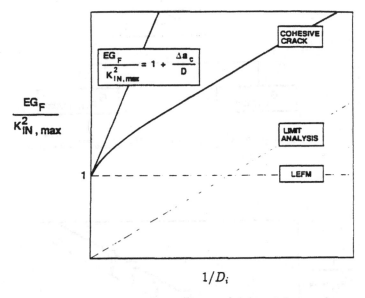

Figure 14. A size-effect plot for the maximum load and first order asymptotic approximation.

dissipation (in other than the cohesive zone) is not included in $G_{F,mod,exp}$, as conceptually illustrated in Fig. 15.

7.2 Size-effect from TPM, SEL and FCM

We are now in a position to compare the maximum load size effects of various cohesive models.

The methodology to compare different cohesive crack models was proposed by Planas and Elices (Planas and Elices, 1988b, c; Llorca et al., 1989) and is based on the effect of the specimen size on the maximum load, with allowance for the fact that the specimens are usually limited in size and that the experimental scatter is high for concrete specimens.

Given a set of models, a set of geometries and a range of practical test specimen sizes, the procedure is as follows:

1. Select one of the models as the *reference model*.

    In this contribution we selected the Hillerborg cohesive crack model with quasi-exponential softening, as the reference. This softening curve better approximates the observed behaviour of concrete.

2. Select one of the geometries as the *reference geometry*.

39

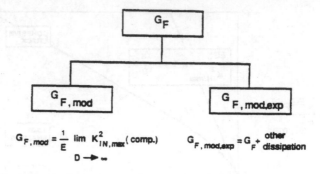

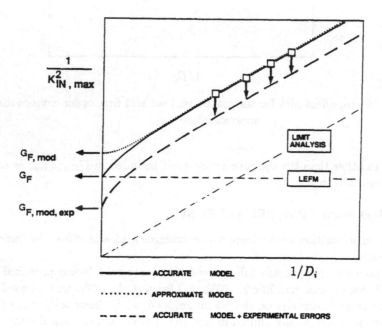

Figure 15: Specific energy concepts; G_F, $G_{F,mod}$, $G_{F,mod,exp}$.

The reference Geometry is taken to be the notched beam, as sketched in Fig. 16.

3. Compute the *reference size effect* for the chosen model in a suitable range of sizes of test specimens.

In this research this range was chosen $100\text{mm} \leq D \leq 400\text{mm}$, where D is the specimen depth.

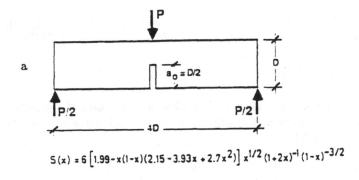

a

$$S(x) = 6\left[1.99 - x(1-x)(2.15 - 3.93x + 2.7x^2)\right] x^{1/2} (1+2x)^{-1} (1-x)^{-3/2}$$

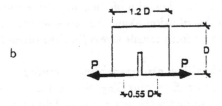

b

$$S(x) = \left[0.886 + 4.64x - 13.32x^2 + 14.72x^3 - 5.6x^4\right] (2+x)(1-x)^{-3/2}$$

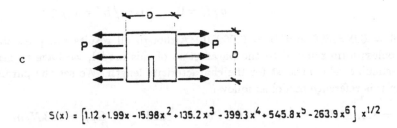

c

$$S(x) = \left[1.12 + 1.99x - 15.98x^2 + 135.2x^3 - 399.3x^4 + 545.8x^5 - 263.9x^6\right] x^{1/2}$$

Figure 16. Selected specimen geometries to explore the influence of stress distribution.

4. Adjust the parameters of the remaining models so that their size effect curves fit well with the reference size effect.

   This adjustment procedure depends on the model and number of parameters to be optimized. It is customary to minimize some error function defined on the experimental interval.

5. Without altering the model parameters determined in step 4, make size effect predictions for the remaining sizes and geometries.

41

Particularly important are extrapolations to large sizes (asymptotic analysis) for which equation (42) may be written specifically as:

$$EG_{F,mod}(K_{INmax})^{-2} = 1 + \Delta a_{C\infty,mod}\, D_i^{-1} + 0\left(\frac{1}{D_i}\right) \qquad (43)$$

Application of this equation will allow a correlation of the values of fracture energy and of critical effective crack extension for the different models.

6. Decide, by inspection of results, whether or not it is possible to discriminate between the different models when experimental scatter is taken into account.

We applied the above procedure to the previously mentioned models and the three selected geometries shown in Fig. 16. The aim of choosing three geometries was to explore the influence of stress distribution along the cross-section, ranging from bending (in the notched beam) to uniform tensile stress (in the notched strut).

7.2.1 Reference model: Cohesive crack with quasi-exponential softening

The essential parameters of the model are E, f_t and G_{FE}, where suffix E in G_{FE} refers to Exponential softening. The assumed softening curve is defined as follows in terms of the stress transferred through the crack faces, σ, and the crack opening w:

$$\sigma/f_t = (1+A)\,exp(-B\,f_t\,w/G_{FE}) - A \; ; \; 0 < w \le 5\,G_{FE}/f_t\,(= w_c)$$
$$\sigma/f_t = 0 \; ; \; 5\,G_{FE}/f_t\,(= w_c) \le w \qquad (44)$$

where $A = 0.0082896$ and $B = 0.96020$. Although all results are presented in dimensionless form referred to the parameters of this model, we have to assume some particular value at least for the characteristic length. We set the parameter values of this reference model as follows:

$$f_t = 3.21\,MPa, \qquad E = 30.0\,GPa \qquad \text{and} \qquad G_{FE} = 103\,N/m$$

from which a characteristic length $l_{chE} = 0.3m$ is obtained. The infinite size critical effective crack extension, $\Delta a_{C\infty E}$, for this particular softening curve, was computed to be $\Delta a_{C\infty E}/l_{chE} = 2.4805$.

The size effect for small specimen sizes was computed by an influence method with a numerical program similar to that used by Petersson (1981), but with a modified algorithm to allow for non-polygonal softening and to speed up convergence. Details are given elsewhere (Elices and Planas, 1990).

The size effect results for the notched beam geometry in the range $0.2 \le a_0/D \le 0.5$ when plotted as l_{chE}/D_i against $EG_{FE}(K_{INmax})^{-2}$ were found (Planas and Elices, 1988b) to be independent of a_0/D to within ± 1 per cent. Hence for this model and for the range of crack depths under study, it appears that the size-effect

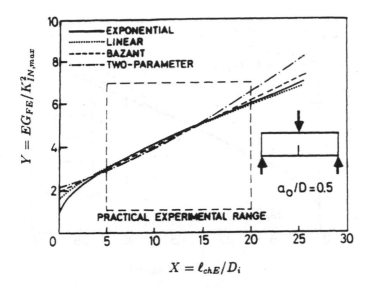

Figure 17. Size-effect results for notched beams

depends only on the intrinsic size, and not on the other geometrical parameters. The average size effect curve is shown in Fig. 17.

The size-effect results for the compact (CT) and single-edge-notched (SEN) specimens for $0.3 \leq a_0/D \leq 0.5$, were also computed. Again, a small dependence on a_0/D was found and results for $a_0/D = 0.5$ are plotted in Figs 18 and 19 for CT and SEN specimens respectively.

7.2.2 Cohesive crack model with linear softening

The essential parameters of the model are E, f_t and G_{FL}, where suffix L in G_{FL} refers to Linear softening. The equation of the softening curve is:

$$\sigma/f_t = 1 - (f_t w/2G_{FL}) \quad ; \quad 0 < w \leq 2\,G_{FL}/f_t (= w_c)$$
$$\sigma/f_t = 0 \quad ; \quad 2\,G_{FL}/f_t (= w_c) \leq w \tag{45}$$

It is assumed that both elastic modulus and tensile strength are the same as for the reference model, so that only one degree of freedom is allowed for curve fitting. In particular, we have for the characteristic length $l_{chL} = (G_{FL}/G_{FE})\,l_{chE}$. The infinite size critical effective crack extension for linear softening, $\Delta a_{C\infty L}$, was computed to be $\Delta a_{C\infty L}/l_{chL} = 0.4195$.

The size-effect curves were computed following exactly the same procedure as for the reference model, and the ratio G_{FL}/G_{FE} was adjusted using graphical

43

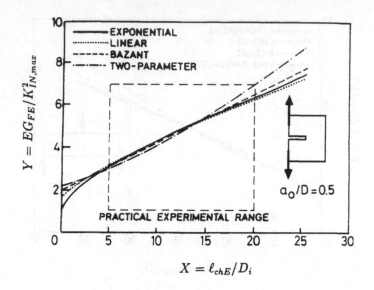

Figure 18: Size-effect results for compact specimens

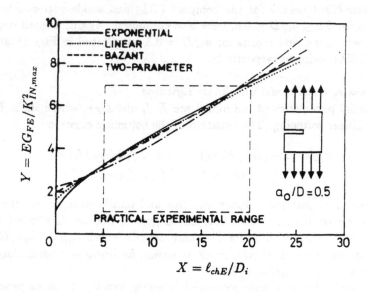

Figure 19: Size-effect results for notched struts

methods to get a good fit with size effect curves for the reference model over the practical test specimen size range.

The results are presented for the three chosen geometries in Figs. 17, 18 and 19, respectively. Taking into account the asymptotic formula (43), it follows that for this model

$$Y = \frac{G_{FE}}{G_{FL}}\left(1 + \frac{\Delta a_{C\infty L}}{l_{chE}}X\right) \tag{46}$$

Again, it was found that the size effect curves are nearly independent of a_0/D, so that the Fig. 17, 18 and 19 are drawn for $a_0/D = 0.5$.

7.2.3 Size-effect law
Bažant's law may be written as

$$EG_{FB}(K_{INmax})^{-2} = 1 + l_0 D^{-1} \tag{47}$$

where G_{FB} is the fracture energy (suffix B refers to Bažant model), and l_0 is a parameter depending *on both geometry and material*. This equation may be rewritten as

$$EG_{FB}(K_{INmax})^{-2} = 1 + \Delta a_{C\infty B}\, D_i^{-1} \tag{48}$$

which is an expression similar to Eq (43) except that $\Delta a_{C\infty B}$ is allowed to depend on geometry. In X,Y coordinates, it reads:

$$Y = \frac{G_{FE}}{G_{FB}}\left(1 + \frac{\Delta a_{C\infty B}}{l_{chE}}X\right) \tag{49}$$

The parameters appearing in Eqn (49) that gave a good fit of Bažant's equation with the reference size-effect were obtained by linear regression. For SEN bending, $a_0/D = 0.5$. A value $G_{FE}/G_{FB} = 1.92$ was obtained for fracture energies, while for the critical effective crack extension $\Delta a_{C\infty B}$ values in the range 0.03 to 0.04m were obtained, depending on the precise geometry. The resulting size-effect straight lines for the three geometries and for $a_0/D = 0.5$ are plotted in Figs 17, 18 and 19.

7.2.4 Two-parameter model
In the two-parameter model it is assumed that in a precracked structure a slow crack growth takes place under increasing load up to a certain crack extension Δa_C at which the maximum load is attained. This critical situation is assumed to occur when the stress intensity factor at the extended crack tip takes the critical value K_{Ic}^s and, simultaneously, the crack opening at the initial crack tip, or $CTOD$, takes its critical value $CTOD_c$.

The asymptotic analysis performed by the authors (Planas and Elices, 1988b,c) lead to the following relationships between the fracture energy G_{FS} and the infinite

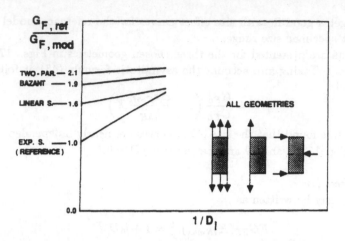

Figure 20. Values of $G_{F,ref}/G_{F,mod}$ for very large sizes. (Reference model; quasi-exponential softening).

size critical effective crack extension $\Delta a_{C\infty S}$ and the primary parameters K^s_{Ic} and $CTOD_c$:

$$K^s_{Ic} = (EG_{FS})^{1/2} \tag{50}$$

$$\Delta a_{C\infty S} = \pi E \frac{CTOD_c^2}{32G_{FS}} \tag{51}$$

such that the size-effect law can again be written as

$$Y = \frac{G_{FE}}{G_{FS}}\left(1 + \frac{\Delta a_{C\infty S}}{l_{chE}}X\right) \tag{52}$$

The determination of the parameters in Eqn (52) that gave a good fit with the reference model is rather cumbersome and will not be reproduced here. For SEN bend specimen, $a_0/D = 0.5$ and the practical size range, the computed parameters were $G_{FE}/G_{FS} = 2.08$ and $\Delta a_{C\infty S} = 0.022m$. Once again little dependence on the initial notch depth was observed (except for very small sizes). The size-effect curves for the three geometries and for $a_0/D = 0.5$ are plotted in Figs 17, 18 and 19.

7.2.5 Behaviour of TPM, SEL and FCM at large sizes

The results in Figs 17, 18 and 19 show that the size effect predictions of the four models for the three geometries differ by less than 5% in the experimental size range. Such small scatter for concrete makes it very difficult to distinguish between these models on the basis of the maximum load criterion.

To reveal differences between the models using this method, one needs to test specimens with very large sizes, roughly an order of magnitude larger than those considered in this work. Despite the good agreement between the models in the small size range, their size-effect curves show clear differences at very large sizes. The two-parameter model, the size-effect law and the fictitious crack model with linear softening, are conservative by 31, 28 and 20 per cent, respectively, when compared with the fictitious crack model with quasi-exponential softening. This result is sketched in Fig. 20 which magnifies the results of Figs 17, 18 and 19 near their origins. A summary of pertinent results is given in Table 11.

7.3 Experimental determination of model parameters

7.3.1 Test programme

As mentioned earlier it is not always possible to determine the essential parameters of all models from the same set of test data. Moreover, since the recommended test procedures are themselves dictated by the models, the same set of test data can, and often does, lead to very different values of the computed fracture energy.

In order to determine (even to over-determine) parameters of TPM, SEL and FCM from the same test data, a special experimental programme was devised. The experimental results were first used to calculate the parameters of the respective models as recommended by their authors: total work of fracture in the RILEM (FMC 50) method, maximum loads in the size-effect law, and peak load and unloading compliance in the two-parameter model. Then for all models a second set of parameters was calculated by the unified maximum-load size-effect methodology described above.

A RILEM (FMC 50) standard concrete was prepared using siliceous rounded gravel (maximum size 10mm) with 400 kg/m^3 of rapid hardening Portland cement and river sand. The mix proportions were 1 : 1.35 : 3.02 : 0.55 (cement : gravel : sand : water) and resulted in a slump of 50mm. Representative concrete properties are listed in Table 12.

The specimens were of the three point bend type of four sizes, as depicted in Fig. 21, but of constant thickness B=100mm. The notch-to-depth ratio of 0.33 does fit in the range for which closed form solutions exist for the essential elastic solutions. However, the span-to-depth ratio of 2.5 (dictated by practical limitations of the frame used for a broader experimental programme) is not one for which closed form solutions exist, and a good deal of work was necessary to determine the functions required for the analysis of experimental results. The central idea was to obtain a rough solution by a combination of the known solutions for pure bending and three point bending (span-to-depth=4) and to refine it by numerical computation, using FEFAP, ANSYS, and weight functions for the notched infinite strip. Details of this procedure will appear elsewhere.

The experimental arrangement permitted measurement of the central force F, the beam deflection Δ and the CMOD, as depicted in Fig. 22. The deflection

Table 11. Summary of results for the different models (all geometries)

Model	Size range (cm)	Max. difference[**] in fracture load (%)	$G_{F,mod}/G_{FE}$
Linear	10 - 40	±2.0	0.64[*]
softening	∞	-20.0	0.64[*]
Size-effect	10 - 40	±2.5	0.52[*]
law	∞	-28.1	0.52[*]
Two-parameter	10 - 40	±4.3	0.48[*]
Model	∞	-30.8	0.48[*]

[*] Fit for SEN-bending, $a_0/D = 0.5$, and practical size range
[**] Relative to FCM with quasi-exponential softening.

was measured relative to the upper side of the specimens in order to exclude any inelastic displacements due to crushing at the supports. The self-weight was compensated by means of prestressed springs, so that the full $F - \Delta$ and $F - CMOD$ curves could be obtained (for a detailed discussion of the test method, see Planas and Elices, 1988a).

The test was run in CMOD control mode to achieve stability. The rate of CMOD was set proportional to the beam depth, and was such that all the sizes went through the maximum load in 30 to 60 seconds, as recommended by the RILEM FMC 50 method (RILEM, 1985).

Table 12. Concrete properties

28 day Tangent Modulus (GPa)	28 day Strength Compressive (MPa)	Splitting (MPa)	Direct Tensile Strength (MPa)	Modulus (GPa)
26.6	33.1	2.8	3.14	25.4

To allow for possible energy dissipation at the supports which can be appreciable, calibration tests were conducted in accordance with the procedure proposed by Elices and Planas (1986a, 1988a). The procedure requires recording of the $F - \Delta$ curve up to the peak load (F_{max}) and of the unloading curve from F_{max} (Fig. 23b). Repetition of the procedure at various values of F_{max} results in the calibration curve for dissipated energy (Fig. 23c).

7.3.2 Fracture parameters following existing procedures
In the RILEM FMC 50 procedure (Hillerborg, 1985; RILEM, 1985) a cohesive crack model with negligible bulk dissipation is postulated, and the external work supplied to rupture the specimen per unit ligament area is set equal to G_F.

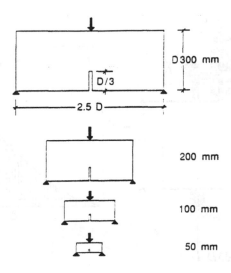

Figure 21: Specimens used in the experimental program.

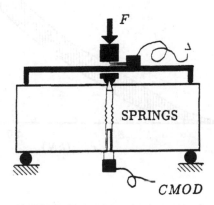

Figure 22: Measurement of force, beam deflection and CMOD

In practice, the energy dissipation at the supports must be subtracted from the total work supplied. The calibration curve in Fig. 23c is the basis for this correction. In our case, this energy dissipation reached almost 12%.

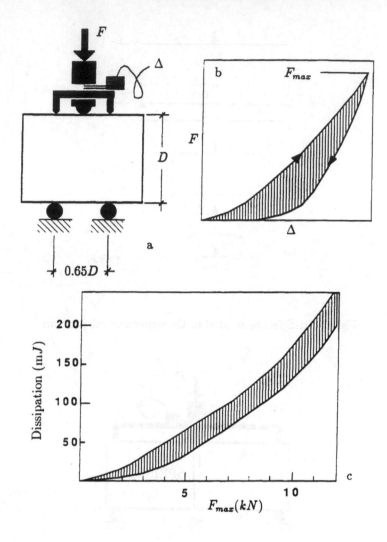

Figure 23: Calibration procedure for energy dissipation at the supports

It is common knowledge that the values of the average specific fracture energy obtained by this procedure depend on the size of the specimen (see, e.g. Hillerborg, 1985). This was also the case for our results; the average specific energy for the largest specimens being about 60% larger than for the smallest specimens.

To account for this scale-dependence of the average fracture energy, the so-called perturbed ligament model (PLM) has been proposed by Elices and Planas. This model rests upon the assumption that at the start of the test the unbroken ligament is already perturbed because of aggregate segregation, or due to shrinkage and thermal effects. Under the simplified assumption that the perturbed zone has a depth independent of the size of the specimen (if it is not too small) one gets a linear relationship between the fracture work per unit thickness and the initial ligament length (Planas and Elices, 1988a). The slope of this line, is by hypothesis, the fracture energy of an eventual unperturbed specimen.

This model has been applied to the present experimental results. In Fig. 24 the fracture work per unit thickness is plotted against the ligament length, and a straight line fitted to the results. The slope of this line is the fracture energy, G_{FPL}, predicted by the PLM, which is compared with the values obtained by the RILEM FMC 50 method, $G_{F,RILEM}$.

Fig 25 shows the test values of $G_{F,RILEM}$ and the PLM value. Even allowing for the inevitable experimental scatter, the size dependence of $G_{F,RILEM}$ is evident, whereas G_{FPL} is constant. In the latter model, the size dependency is attributed to specimen manufacturing and handling condition. Also shown on the figure is the fracture energy determined according to the size-effect law, which turns out to be $G_{FB} \approx 40\,N/m$.

The results according to the two-parameter model showed that both the G_F and the $CTOD_c$ are size dependent, but it is likely that this is a result of experimental inaccuracies (Shah, 1990). No unloading past the peak load at 95% of P_{max} was performed as required by the TPM in the determination of the unloading compliance C_u and therefore of the effective Griffith crack length a (see Section 3).

7.3.3 Fracture parameters following size effect fitting

We may exploit the fact that the size effect curves are quite geometry-insensitive when plotted against the intrinsic size. At first glance it may appear that this curve fitting process is very complex. This is indeed so if an analytical approach is taken. But there is a graphical approach which is quick, albeit somewhat approximate. In this approach dimensional plots $(K_{INmax})^{-2}$ versus $1/D$ are obtained from the dimensionless plots in Fig. 17 by scale changes, or, in geometrical terms, by a simple affinity. The scale changes are directly related to the parameters of the model. On a log-log plot the affinity (change of scale) is transformed into a translation vector, the components of which are related to the scale changes, and hence, to the material parameters. This simple graphical approach consists of the following steps.

1. Redraw the size effect curve (Fig. 17) for the chosen model on a log-log tracing paper;

2. Draw the experimental results $(K_{INmax})^{-2}$ versus $1/D_i$ on a log-log plot on another sheet of paper.

3. Slide the tracing paper over the experimental record until a good fit is obtained (use of a parallel drawing machine simplifies the procedure);

51

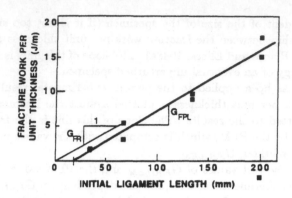

Figure 24: Determination of the specific fracture energy according to the Perturbed Ligament Model: G_{FPL}.

4. Determine the translation vector graphically. The components of the translation vector are $\{\lg(l_{chE}), \lg(EG_{FE})\}$ from which l_{chE} and G_{FE} can be obtained. These in turn are related to the parameters of the chosen model through the values given in Table 11.

This procedure is valid for models with two degrees of freedom, such as the size-effect law and the two-parameter model. But in cohesive cracks, where E and f_t are independently measured, there is only one degree of freedom, namely G_F, and the two components of the translation vector are no longer independent. In fact the translation is decomposed into a fixed known translation with components $\{\lg(f_t^{-2}E), \lg E\}$ followed by a translation parallel to the $\{1, 1\}$ direction: $\{\lg G_{FE}, \lg G_{FE}\}$. This method was applied to the experimental results (for the size-effect law a standard linear regression was used), and the results are shown in Fig. 26 together with the values of the parameters. It is clear that all the mo els fit the experimental results within 10%, but that the fracture energies at infinite size are different by a factor of two.

7.4 Discussion of results

In this contribution the essential points regarding the size effect behaviour of maximum load are analyzed with particular emphasis on the simple asymptotic structure valid for most reasonable fracture models.

The asymptotic structure allowed a model-independent definition of the fracture energy although the results based on practical (small size) specimens will be model-dependent and will represent a model parameter rather than a material property.

A recently proposed theoretical procedure for a unified comparison of fracture models was summarized and applied to different models and geometries. The essential conclusion is that for these geometries, and for the usual range of sizes

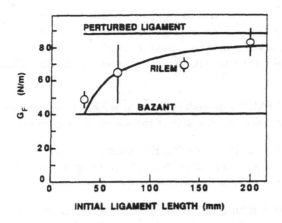

Figure 25: Specific fracture energy according to RILEM, SEL and perturbed ligament models ($G_{F,RILEM}, G_{FB}, G_{FPL}$).

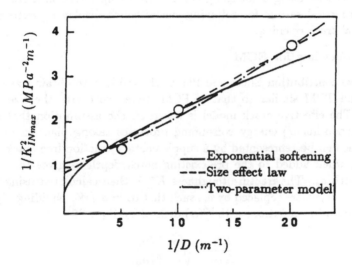

Figure 26: Size-effect curves for various models obtained by parameter fitting.

tested in the laboratory, all models will describe the size effect within experimental scatter, but that extrapolation to large sizes will give differing results.

The results of a series of stable tests on notched beams of four sizes were analyzed following different procedures. From these results the following conclusions may be drawn:

1. The RILEM FMC50 procedure leads to values of the fracture energy that are strongly dependent on the size of the specimen (especially for D <150 mm).

2. The size dependency of the RILEM fracture energy value is well explained by the perturbed ligament model. The latter gives the fracture energy corresponding to a virtual unperturbed specimen.

3. The size effect fitting procedure results in model parameters which are in good agreement with the theoretically predicted parameters.

4. The size effect fitted value of the fracture energy for SEL and FCM with linear softening is much less than that given by perturbed ligament extrapolation.

5. The FCM with quasi-exponential softening gives a value of fracture energy within 13% of the perturbed ligament extrapolation. Agreement could be enhanced by using a softening curve with a longer tail, in accordance with experimental results for softening curves which display an extremely large critical crack opening.

7.5 Size effect law for ECM

In the above contribution Elices and Planas showed how to obtain size effect laws for SEL and TPM similar to that of FCM. Here, we derive the size effect law for ECM. The effective crack model is based on the assumption that the effect of various (non-linear) energy consuming processes taking place in the fracture process zone can be represented by a supplementary traction-free crack (Fig. 28). The latter, when added to the pre-existing notch depth a_0 gives the size of the effective notch a_e. The fracture toughness K_{Ic}^e is then calculated using LEFM in which a_0 is everywhere replaced by a_e, such that $\alpha_e = a_e/W$. Dividing K_{Ic}^e by K_{IN} yields the following equation, with $F(\alpha)$ defined by Eqn. 26

$$\frac{K_{Ic}^e}{K_{IN}} = \sqrt{\frac{\alpha_e}{\alpha_0} \frac{F(\alpha_e)}{F(\alpha_0)}}. \tag{53}$$

According to ECM, the effective critical stress intensity factor K_{Ic}^e is a material constant; thus the size effect is embedded in a_e or, more accurately, in $\Delta a_e = (a_e - a_0)$.

An empirical relation for a_e is available based on extensive test data on laboratory-size three point bend specimens (see Eqn. 31, Section 5.3). It should be emphasized

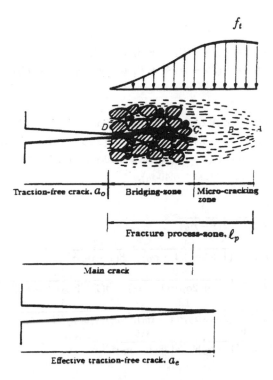

Figure 27: Schematic illustration of effective notch depth a_e and process zone.

that this relation is valid only in the range of depths $100 \leq W \leq 400$ (mm). It cannot therefore be used for extrapolation to $W \to \infty$.

It is possible to express Eqn. 53 as a series in descending powers of W by expanding its RHS in powers of $\Delta a_e/W$ and retaining only terms up to and including the desired order (say, four)

$$\left(\frac{K_{Ic}^e}{K_{IN}}\right)^2 = 1 + \lambda_1\left(\frac{\Delta a_e}{W}\right) + \lambda_2\left(\frac{\Delta a_e}{W}\right)^2 + \lambda_3\left(\frac{\Delta a_e}{W}\right)^3 + \lambda_4\left(\frac{\Delta a_e}{W}\right)^4, \quad (54)$$

where the dimensionless coefficients $\lambda_1, ..., \lambda_4$ depend only on $\alpha_0 = a_0/W$

$$
\begin{aligned}
\lambda_1 &= (2D_1 + 1/\alpha_0), \\
\lambda_2 &= (D_1^2 + 2D_2) + (2D_1)/\alpha_0, \\
\lambda_3 &= 2(D_3 + D_1 D_2) + (D_1^2 + 2D_2)/\alpha_0, \\
\lambda_4 &= (D_2^2 + 2D_1 D_3 + 2D_4) + 2(D_3 + D_1 D_2)/\alpha_0.
\end{aligned}
\quad (55)
$$

Here

$$D_1 = (B_1 + B_0 A_1)/B_0,$$

$$D_2 = (B_2 + B_1 A_1 + B_0 A_2)/B_0,$$
$$D_3 = (B_3 + B_2 A_1 + B_1 A_2 + B_0 A_3)/B_0, \tag{56}$$
$$D_4 = (B_4 + B_3 A_1 + B_2 A_2 + B_1 A_3 + B_0 A_4)/B_0,$$

where

$$B_0 = 1.99 - \alpha_0(1 - \alpha_0)(2.15 - 3.93\alpha_0 + 2.70\alpha_0^2),$$
$$B_1 = \alpha_0(1 - \alpha_0)(3.93 - 5.40\alpha_0) - (1 - 2\alpha_0)(2.15 - 3.93\alpha_0 + 2.70\alpha_0^2),$$
$$B_2 = 2.70\alpha_0(1 - \alpha_0) + (1 - 2\alpha_0)(3.93 - 5.40\alpha_0) + (2.15 - 3.93\alpha_0 + 2.70\alpha_0^2),$$
$$B_3 = 2.70(1 - 2\alpha_0) - (3.93 - 5.40\alpha_0),$$
$$B_4 = -2.70,$$
$$A_1 = -\frac{2}{(1 + 2\alpha_0)} + \frac{3}{2(1 - \alpha_0)}, \tag{57}$$
$$A_2 = \frac{4}{(1 + 2\alpha_0)^2} - \frac{3}{(1 + 2\alpha_0)(1 - \alpha_0)} + \frac{15}{8(1 - \alpha_0)^2},$$
$$A_3 = -\frac{8}{(1 + 2\alpha_0)^3} + \frac{6}{(1 + 2\alpha_0)^2(1 - \alpha_0)} - \frac{15}{4(1 + 2\alpha_0)(1 - \alpha_0)^2} + \frac{35}{16(1 - \alpha_0)^3},$$
$$A_4 = \frac{16}{(1 + 2\alpha_0)^4} - \frac{12}{(1 + 2\alpha_0)^3(1 - \alpha_0)} + \frac{15}{2(1 + 2\alpha_0)^2(1 - \alpha_0)^2}$$
$$- \frac{35}{8(1 + 2\alpha_0)(1 - \alpha_0)^3} + \frac{315}{128(1 - \alpha_0)^4}.$$

Formula 54 which is similar in form to Eqn. 38 describes the size effect for ECM and reduces to LEFM result ($K_{IN} \to K_{Ic}^e = K_{Ic}$) when $W \to \infty$. However, comparison with the size effect for FCM (Eqn. 38) is facilitated if Eqn. 54 is further transformed to read

$$\frac{EG_{FE}}{K_{IN}^2} = \frac{G_{FE}}{G_F^e} \left[1 + \gamma_1\left(\frac{l_{ch}}{W}\right) + \gamma_2\left(\frac{l_{ch}}{W}\right)^2 + \gamma_3\left(\frac{l_{ch}}{W}\right)^3 + \gamma_4\left(\frac{l_{ch}}{W}\right)^4\right], \tag{58}$$

where the toughness G_F^e corresponding to K_{Ic}^e is a material constant

$$G_F^e = \frac{(K_{Ic}^e)^2}{E}, \tag{59}$$

and

$$\gamma_1 = \lambda_1(\Delta a_e/l_{ch}),$$
$$\gamma_2 = \lambda_2(\Delta a_e/l_{ch})^2,$$
$$\gamma_3 = \lambda_3(\Delta a_e/l_{ch})^3, \tag{60}$$
$$\gamma_4 = \lambda_4(\Delta a_e/l_{ch})^4.$$

7.6 Comparison of size effects for FCM and ECM

In order to compare the size effect for ECM with that of FCM we follow the methodology of Elices and Planas and consider practical notched test beam sizes in the range $100 - 400$mm with a_0/W in the range $0.2 - 0.5$. The method adopted by Elices and Planas takes the FCM as the main input and attempts to determine the parameters of ECM in such a way as to achieve the best fit with the size effect for FCM over the practical size range.

7.6.1 Material properties and calculation of critical load

We use the mix parameters of Elices and Planas, namely $f_t = 3.21 MPa, E = 30.0 GPa, G_{FE} = 103 N/m$ and $S/W = 4$. Moreover, for FCM we again use their quasi-exponential softening curve (Eqn. 44). For each beam specimen finite element calculations were carried out in conformity with FCM to obtain the maximum fracture load. K_{IN} corresponding to this load was calculated, thereby completely determining the LHS of the size effect law Eqn. 38 for FCM. It is now possible to plot the size effect for FCM. This has been done in Fig. 28 for two values of a_0/W. As observed by Elices and Planas, the variation with a_0/W is not significant.

7.6.2 Size effect plots for ECM

The size effect law (Eqn. 58) for ECM can be similarly depicted provided we know G_F^e and $\gamma_1, ..., \gamma_4$ in Eqn. 58. For the above beam specimens G_F^e was calculated using Eqn. 59 in which a_e was calculated from the regression formula (Eqn. 31). For the purposes of this exercise g was set equal to 20mm in this formula. It should be noted that a_e varies only slightly with $g \geq 10$mm. The values of a_e determined from this formula and other material properties are given in Table 13.

The mean value of G_F^e was regarded as a material constant for the mix. It should be noted that the sizes under consideration, viz. $100 \leq W \leq 400$mm are within the range of validity of the regression equation for a_e. The constants $\gamma_1, ..., \gamma_4$ (Eqn. 60) can therefore be defined only in this range which corresponds to $0.75 \leq l_{ch}/W \leq 3.0$. This limits the scope of comparison with FCM. For a meaningful comparison with FCM in the full range of l_{ch}/W, we followed an inverse procedure, whereby $\Delta a_e/l_{ch}$ for each specimen was calculated by solving a fourth order equation in $\Delta a_e/l_{ch}$. This equation was obtained from Eqns. 58 and 60 by requiring complete agreement between the size effects for ECM and FCM over the practical test specimen size range under consideration. This inverse procedure allows us to calculate the coefficients $\gamma_1, ..., \gamma_4$ appearing in Eqn. 58 that are valid for the full range of l_{ch}/W considered in Fig. 28. Of course, this inverse procedure is only valid if the values of $\Delta a_e/l_{ch}$, and therefore of a_e/W, so calculated are in good agreement with a_e/W calculated using the regression formula in the range of validity of the latter, i.e. $100 \leq W \leq 400$mm. As is clear from Fig. 29 and Table 13 (columns 3 and 4), the agreement is indeed very good.

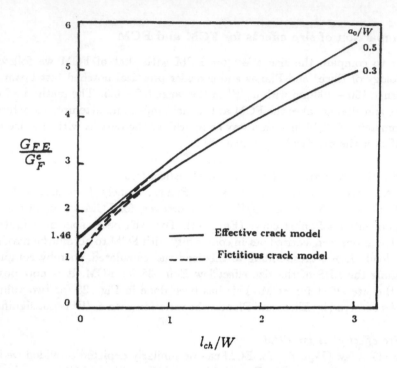

Figure 28: Comparison of size effect plots predicted by ECM with those predicted by FCM.

The size effect law predicted by ECM is exhibited on Fig. 28 for two values of a_0/W. For ease of using the size effect law (Eqn. 58), Table 14 gives the values of various parameters and comparison with the fracture load calculated according to FCM. It is clear that the predictions of the two models differ by no more than about 17 per cent for very large sizes $(W \to \infty)$.

It should be noted that whereas $\lambda_1, ..., \lambda_4$ vary only with a_0/W, the coefficients $\gamma_1, ..., \gamma_4$ just as $\Delta a_e/l_{ch}$ vary with both W and a_0/W. However, $\Delta a_e/l_{ch}$ should not depend on size, if it is to be a material property. In fact $\Delta a_e/l_{ch}$ is independent of size beyond a certain depth of the specimen as is evident in Fig. 30; it exibits an asymptotic behaviour with increasing W/l_{ch} and attains a constant value, which in our case is at a depth of around 500mm.

8 Acknowledgement

It is pleasant duty to thank all members of Sub-Committee A for their co-operation in the preparation of this report. The extensive comments on the several draft versions of the report from Professors Swartz, Shah and Kasperkiewicz, and Dr. Hassanzadeh were most valuable in improving both the content and presentation of the report. The companion contribution by Professors Elices and Planas was

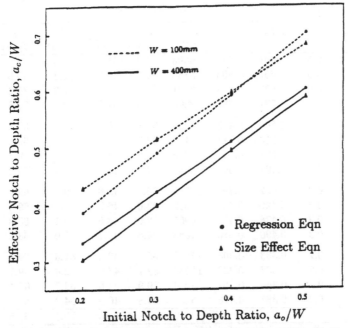

Figure 29: Comparison of effective notch depth ratio a_e/W calculated using the regression equation 31, and the size effect equation 58.

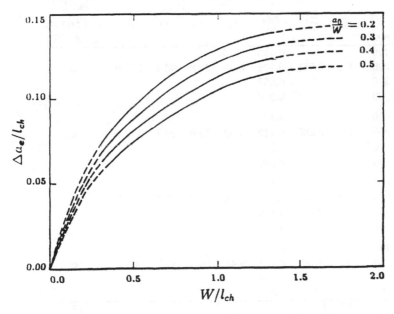

Figure 30: Variation of $\Delta a_e/\ell_{ch}$ for laboratory size specimens.

Table 13. Specimen details and material properties

W (mm)	a_0/W	a_e/W	a_e^*/W	P_{max} (N)	K_{IN}	K_{Ic}^e $MPa\sqrt{mm}$	G_F^e (J/m)	$G_F^{e\dagger}$ (J/m)
400	0.5	0.601	0.586	6393	34.0	48.4	78.0	69.7
300	0.5	0.607	0.605	5158	31.7	46.2	71.3	69.9
200	0.5	0.626	0.631	3801	28.6	44.9	67.3	70.2
100	0.5	0.701	0.680	2203	23.5	51.8	89.4	72.8
400	0.4	0.509	0.493	8784	34.8	48.1	77.0	69.8
300	0.4	0.513	0.511	7158	32.7	45.9	70.2	69.3
200	0.4	0.528	0.541	5286	29.6	43.7	63.6	69.2
100	0.4	0.591	0.596	3085	24.4	45.0	67.4	70.0
400	0.3	0.422	0.398	11585	35.2	48.8	79.4	69.5
300	0.3	0.426	0.421	9393	33.0	46.2	71.2	69.3
200	0.3	0.438	0.453	6965	30.0	43.5	63.1	68.8
100	0.3	0.490	0.513	4083	24.8	42.1	59.0	68.5
400	0.2	0.333	0.303	14892	35.0	49.4	81.3	69.5
300	0.2	0.337	0.328	12062	32.7	46.6	72.4	69.1
200	0.2	0.346	0.361	8984	29.9	43.6	63.3	68.4
100	0.2	0.387	0.428	5267	24.8	40.3	54.0	68.0
						Mean	70.5	69.5

* – size effect prediction of a_e/W † – G_F^e calculated with a_e^*/W

Table 14. Size effect parameters for ECM

W (mm)	$\frac{a_0}{W}$	$\frac{G_F^e}{G_{FE}}$	$\frac{\Delta a_e}{l_{ch}}$	λ_1	λ_2	λ_3	λ_4	Max Diff in Fracture Loads (%)
400			0.115					
300	0.5	0.685	0.105	6.35	26.8	93.4	283.0	0
200			0.087					
100			0.060					
400			0.124					
300	0.4	0.685	0.112	5.53	19.9	60.7	152.0	0
200			0.094					
100			0.066					
400			0.131					
300	0.3	0.685	0.121	5.12	15.3	46.1	91.8	0
200			0.102					
100			0.071					
400			0.138					
300	0.2	0.685	0.128	5.41	10.6	45.7	47.9	0
200			0.108					
100			0.076					
∞	All	0.685						-17.3

completed during the senior author's (BLK) visit to Madrid. He thanks them for making the visit possible and for their generous hospitality. Their contribution was prepared with the help of G.V. Guinea under a financial grant (PB-86-0494) from the U.S-Spain Joint Committee for Science and Technology.

9. References

Alexander, M., (1987) Data from tests on notched concrete beams, *Private Communication*, University of Witwatersrand, Johannesburg.

Bascoul, A., (1987) *Private Communication*, University of Paul-Sabatier, Toulouse.

Bažant, Z.P., (1984) Size effect in blunt fracture: concrete, rock, metal, *J. of Eng. Mech., ASCE*, V110, pp. 518-535.

Bažant, Z.P., (1987) Determination of fracture energy of heterogeneous material and similitude, *SEM-RILEM Int. Conf. on Fracture of Concrete and Rock*, Ed: S.P. Shah and S. Swartz, Houston, pp. 390-402.

Bažant, Z.P., Kim, J.K. and Pfeiffer, P.A., (1986) Determination of fracture properties from size effect tests, *J. of Struct. Eng., ASCE*, V112, No.2, pp. 289-307.

Bažant, Z.P. and Pfeiffer, P.A., (1986) Determination of fracture energy from size effect and brittleness number, *Report No. 86-8/428d*, Northwestern University.

Bažant, Z.P. and Pfeiffer, P.A., (1987) Determination of fracture energy from size effect and brittleness number, *ACI Materials J.*, V84.

Brown, W.F. and Srawley, J.E., (1966) Crack toughness testing of high strength matallic materials, *ASTM-STP No. 410*, p. 130.

Carpinteri, A., (1980) Experimental determination of fracture toughness parameters K_{Ic} and J_{Ic} for aggregative materials. In *Advances in Fracture Research*, (Francois, D. *et al.*, Eds) Pergamon Press, pp. 1491-1498.

Carpinteri, A., (1984) Numerical modelling of damage and fracture concrete. In *Advances in Fracture Research*, (Valluri, S. R. *et al.*, Eds) Pergamon Press, pp. 2801-2807.

Catalano, D.M. and Ingraffea, A.R., (1982) Concrete fracture: A linear elastic fracture mechanics approach, *Report No. 82-1*, Department of Structural Engineering, Cornell University.

Elices, M., Corres, H. and Planas, J., (1987) Experimental results of fracture energy of concrete for different specimen sizes, *Report to RILEM TC 50-FMC*, Madrid.

Elices, M. and Planas, J. (1989), Material Models, Chap. 3 in *Fracture Mechanics of Concrete Structures: From Theory to Applications*, (Elfgren, L. Ed.), Chapman and Hall, pp. 16-66.

Ferrara, G., (1987) *Private Communication*, Italian Electricity Board (ENEL-DSR-CRIS), Milano.

Hillerborg, A., Modeer, M. and Petersson, P.E. (1976), Analysis of Crack Formation and Crack Growth in Concrete by Means of Fracture Mechanics and Finite Elements, *Cement Concr. Res.*, V6, pp. 773-782.

Hillerborg, A. (1985), The Theoretical Basis of a Method to Determine the Fracture Energy G_F of Concrete, *Materials and Structures*, V18 pp. 291-296.

Hilsdorf, H.K. and Brameshuber, W., (1987) Probengrö Benabhängigkeit bruchmechanischer Kennwarte für Beton (Size dependency of fracture mechanics parameters of concrete), *Private Communication*,.

Horvath, R. and Persson, T., (1984) The influence of the size of the specimen on the fracture energy of concrete, *Report TUBM-5005*, Lund, Sweden, p. 45.

Jenq, Y.S. and Shah, S.P., (1984) Non-linear fracture parameters for cement based composites: theory and experiments, *Proceedings of the NATO Advanced Workshop: Application of fracture mechanics to cementitious composites*, Editor: S.P. Shah, Northwestern University, pp. 213-253.

Jenq, Y.S. and Shah, S.P., (1985) Two parameter fracture model for concrete, *J. of Eng. Mech., ASCE*, 111(10), pp 1227-1241.

John, R. and Shah, S.P., (1987) Effect of high strength and rate of loading on fracture parameters of concrete, *Proceedings, RILEM-SEM International Conference on Fracture of Concrete and Rock*, Editors: Shah, S.P. and Swartz, S.E., Houston, pp. 35-52.

John, R., Shah, S.P. and Jenq, Y.S., (1987) A fracture mechanics model to predict the rate sensitivity of mode I fracture for concrete, *Cem. and Conc. Research*, V17, pp. 249-262.

Karihaloo, B.L. and Nallathambi, P., (1989 a) An improved effective crack model for the determination of Fracture toughness of concrete, *Cem. and Conc. Research*, V19, pp. 603-610.

Karihaloo, B.L. and Nallathambi, P., (1989 b) Fracture toughness of plain concrete from three-point bend specimens, *Matériaux et Constructions*, V22, pp. 185-193.

Karihaloo, B.L. and Nallathambi, P., (1989 c) Effective crack model and tension-softening models, *Fracture of Concrete and Rock: Recent Developments* (Eds: S. P. Shah, S. E. Swartz and B. Barr), Elsevier Science., pp. 701-710.

Kasperkiewicz, J., (1990) *Private Communication,* Institute of Fundamental Tech. Research, Warsaw.

LLorca, J., Planas, J. and Elices M. (1989), On the Use of Maximum Load to Validate or Disprove Models of Concrete Fracture Behaviour, *Fracture of Concrete and Rock: Recent Developments* (Eds: S. P. Shah, S. E. Swartz and B. Barr), Elsevier Science, pp. 357-368.

Malvar, J.L., (1987) *Private Communication,* Naval Civil Engineering Laboratory, California. See also Report from Malvar, October 1987.

Mindess, S., (1984) The effect of specimen size on the fracture energy of concrete, *Cement and Concrete Research,* Vol. 14, pp. 431-436.

Murakami, Y., et al., (1987) Stress intensity factors handbook, *Pergamon Press,* Oxford-NewYork.

Nallathambi, P. (1986) Fracture behaviour of plain concretes, *Doctoral Dissertation,* University of Newcastle, Australia, p. 207.

Nallathambi, P. and Karihaloo, B.L., (1986 a) Influence of slow crack growth on the fracture toughness of plain concrete, *Proceedings of the International Conference on Fracture Toughness and Fracture Energy of Concrete,* Editor: F.H. Wittman, Elsevier Sciences Publications, Amsterdam, pp. 271-280.

Nallathambi, P. and Karihaloo, B.L., (1986 b) Determination of specimen-size independent fracture toughness of plain concrete, *Magazine of Concrete Research,* V38, No. 135, pp. 67-76.

Nallathambi, P. and Karihaloo, B.L., (1986 c) Stress intensity factor and energy release rate for three-point bend specimens, *Engineering Fracture Mechanics,* V25, No. 3, pp. 315-321.

Nallathambi, P. and Karihaloo, B.L., (1990) Fracture of concrete: application of effective crack model, *Proc. of the 9th Int. Conf. on Expt. Mech.,* Lyngby, Denmark.

Petersson, P.E. (1981), Crack Growth and Development of Fracture Zones in Plain Concrete and Similar Materails, *Report TVBM 1006,* University of Lund, Sweden.

Planas, J. and Elices, M., (1986a) Towards a measure of G_F: An analysis of experimental results, *Fracture Toughness and Fracture Energy of Concrete*, (F.H. Wittmann, Ed.), Elsevier Science Publishers, Amsterdam, pp. 381-390.

Planas, J. and Elices, M. (1986b), Un Nuevo Método de Análisis del Comportamiento Asintótico de la Propagación de una Fisura Cohesiva en Modo I, *Anales de Mecánica de la Fractura*, No. 3, pp. 219-227.

Planas, J. and Elices, M. (1987), Asymptotic Analysis of the Development of a Cohesive Crack Zone in Mode I Loading for Arbitrary Softening Curves, *Proceedings SEM-RILEM International Conference*, Houston, pp. 384.

Planas, J. and Elices, M. (1988a), Conceptual and Experimental Problems in the Determination of the Fracture Energy of Concrete, *Proc. Int. Workshop on Fracture Toughness and Fracture Energy*, Sendai, Japan. Also in *Fracture Toughness and Fracture Energy: Test Methods for Concrete and Rock*, (Mihashi, H., et al. Eds.) Balkema, Rotterdam, 1989, pp. 165-181.

Planas, J. and Elices, M. (1988b), Size Effect in Concrete Structures: Mathematical Approximations and Experimental Validation, *Proc. France-USA Workshop on Strain Localization and Size Effect due to Cracking and Damage*, Cachan, Paris, France, September 6-9.

Planas, J. and Elices, M. (1988c), Fracture Criteria for Concrete: Mathematical Approximations and Experimental Validation, *Proc. Int. Conf. on Fracture and Damage of Concrete and Rock*, Vienna, Austria, July 4-6. Also in *Engineering Fracture Mechanics*, Vol. 35, 1990, pp. 87-94.

Planas, J. and Elices, M. (1990), A Nonlinear Analysis of a Cohesive Crack, To be published in *J. Mech. and Phys. of Solids*.

Refai, M.E. and Swartz, S.E., (1987) Fracture behaviour of concrete beams in three-point bending considering the influence of size effects, *Report No. 190*, Engineering Experimental Station, Kansas State University, p. 241.

Shah, S. P., (1990), *Private Communication*, Northwestern University, Evanston, Illinois.

Tada, H., Paris, P.C. and Irwin, G.R., (1973, 85), The Stress analysis of cracks handbook, *Del Research Corp*, Hellertown, PA.

APPENDIX I

This appendix gives details of test data (with source names) used in the analysis according to Proposal 1. The summary of analysis results was given in Table 1.

Table 15. (Refai and Swartz, 1987; Notched beams, $g = 19mm$,)

Serial No.	Beam dimensions, mm			a_0/W	$\underline{a}/W$	K_{Ic}^s ($MPa\sqrt{m}$)	$CTOD_c$ (mm)
	S	B	W				
(Mix A. $w/c = 0.5$, $f_c' = 55.8\,MPa$, $E = 36.8\,GPa$)							
1	381	76	102	0.290	0.550	1.153	0.0219
2	381	76	102	0.320	0.560	1.184	0.0214
3	381	76	102	0.460	0.680	1.158	0.0217
4	381	76	102	0.520	0.690	0.939	0.0084
5	381	76	102	0.620	0.720	0.594	0.0067
6	381	76	102	0.670	0.810	0.560	0.0086
(Mix B. $w/c = 0.5$, $f_c' = 53.1\,MPa$, $E = 38.4\,GPa$)							
7	762	76	203	0.290	0.410	1.054	0.0153
8	762	76	203	0.300	-	-	-
9	762	76	203	0.500	-	-	-
10	762	76	203	0.700	-	-	-
(Mix C. $w/c = 0.5$, $f_c' = 54.4\,MPa$, $E = 39.3\,GPa$)							
11	1143	76	305	0.300	0.420	1.378	0.0240
12	1143	76	305	0.500	0.630	1.250	0.0257
13	1143	76	305	0.670	0.730	0.755	0.0102

Table 16. (Refai and Swartz, 1987; Precracked beams)

Serial No.	Beam dimensions, mm			a_0/W	$\underline{a}/W$	K^s_{Ic}	$CTOD_c$
	S	B	W			$(MPa\sqrt{m})$	(mm)
Mix B. $g = 19mm$, $f'_c = 53.1MPa$, $E = 38.4GPa$)							
1	762	76	203	0.250	0.440	1.142	0.0222
2	762	76	203	0.275	0.480	1.160	0.0243
3	762	76	203	0.293	0.620	1.124	0.0360
4	762	76	203	0.454	0.720	1.156	0.0350
5	762	76	203	0.563	0.780	1.212	0.0337
6	762	76	203	0.608	0.830	1.264	0.0385

Note: Initial crack of depth a_0 was created by load recycling the specimen. There was no noticeable difference in K^s_{Ic} values between notched and precracked beams (cf. Mix B. Table 15).

Table 17. (Refai and Swartz, 1987; Precracked beams)

Serial No.	Beam dimensions, mm			a_0/W	$\underline{a}/W$	K^s_{Ic}	$CTOD_c$
	S	B	W			$(MPa\sqrt{m})$	(mm)
Mix C. $g = 19mm$, $f'_c = 54.4MPa$, $E = 39.3GPa$)							
1	1143	76	305	0.158	0.540	1.251	0.0500
2	1143	76	305	0.187	0.400	1.319	0.0323
3	1143	76	305	0.208	0.550	1.228	0.0457
4	1143	76	305	0.221	0.430	1.224	0.0301
5	1143	76	305	0.250	0.490	1.337	0.0373
6	1143	76	305	0.258	0.460	1.243	0.0305
7	1143	76	305	0.266	0.420	1.094	0.0222
8	1143	76	305	0.411	0.580	1.309	0.0309
9	1143	76	305	0.454	0.600	1.312	0.0285
10	1143	76	305	0.479	0.620	1.138	0.0245
11	1143	76	305	0.500	0.650	1.052	0.0240
12	1143	76	305	0.545	0.710	1.138	0.0289
13	1143	76	305	0.570	0.690	1.033	0.0209

Table 18. (Bascoul, 1987. $g = 3.15mm$)

Serial No.	Beam dimensions, mm			a_0/W	a/W	K^s_{Ic} $(MPa\sqrt{m})$	$CTOD_c$ (mm)
	S	B	W				
(Crushed Marble, $w/c = 0.53$, $f'_c = -$, $E = 36.8\,GPa$[†])							
1	160	24	40	0.250	0.323	0.982	0.0037
2	160	24	39	0.263	0.307	0.854	0.0025
3	160	24	40	0.274	0.301	0.792	0.0018
4	320	50	80	0.512	0.550	0.931	0.0050
5	320	50	80	0.500	0.545	0.914	0.0053
6	320	50	80	0.512	0.562	0.923	0.0057
7	320	49	80	0.262	0.310	1.025	0.0057
8	320	50	83	0.263	0.308	0.899	0.0048
9	320	50	83	0.264	0.310	0.869	0.0047
10	320	50	80	0.125	0.154	0.779	0.0032
11	320	50	83	0.156	0.191	0.843	0.0038
12	320	50	83	0.144	0.176	0.911	0.0039

[†] Estimated from load - deflection plots.

Table 19. (Catalano and Ingraffea, 1982)

Serial No.	Beam dimensions, mm			a_0/W	a/W	K^s_{Ic} $(MPa\sqrt{m})$	$CTOD_c$ (mm)
	S	B	W				
(Mortar, $g = 6mm$, $w/c = 0.44$, $f'_c = 60.7\,MPa$, $E = 33.5\,GPa$)							
1	1321	102	305	0.197	0.278	1.236	0.0104
2	1321	102	305	0.211	0.248	1.046	0.0185
(Conc., $g = 12.7mm$, $1:2:1.62$, $w/c = 0.52$, $f'_c = 45.5\,MPa$, $E = 31.0\,GPa$)							
3	1321	152	305	0.205	0.261	1.286	0.0131
4	1321	152	305	0.198	0.352	1.736	0.0227
5	1321	152	305	0.202	0.252	1.403	0.0302
(Conc., $g = 12.7mm$, $1:2:2.44$, $w/c = 0.52$, $f'_c = 43.4\,MPa$, $E = 31.0\,GPa$)							
6	1321	152	305	0.211	0.267	1.529	0.0196
7	1321	152	305	0.202	0.244	1.504	0.0156
8	1321	152	305	0.205	0.275	1.558	0.0155

Table 20. (Jenq and Shah, 1984; John and Shah, 1987)

Serial No.	Beam dimensions, mm			a_0/W	$\underline{a}/W$	K^s_{Ic} $(MPa\sqrt{m})$	$CTOD_c$ (mm)
	S	B	W				
(Concrete, $g = 19mm, w/c = 0.65$, $f'_c = 25.16\,MPa$, $E = 27.24\,GPa$)							
1	914	86	229	0.333	0.347	1.106	0.0076
2	914	86	229	0.333	0.385	0.940	0.0122
3	609	57	152	0.318	0.377	0.787	0.0089
4	609	57	152	0.324	0.511	1.001	0.0269
5	305	29	31	0.293	0.498	0.968	0.0196
6	305	29	31	0.293	0.548	0.891	0.0180
7	305	29	31	0.293	0.535	0.989	0.0198
8	305	29	31	0.293	0.531	1.127	0.0231
(High st. Concrete, $g = 8mm, w/c = 0.22$, $f'_c = 110 MPa, E = 56.55 GPa$)							
9	203	25	76	0.329	†	2.130	0.0366
10	203	25	76	0.329	†	2.130	0.0310

Notes: † Values not available.

Table 21. (Hilsdorf and Brameshuber, 1987)

Serial No.	Beam dimensions, mm			a_0/W	$\underline{a}/W$	K^s_{Ic} $(MPa\sqrt{m})$	$CTOD_c$ (mm)
	S	B	W				
(Concrete, $g = 32mm, w/c = 0.54$, $f'_c = 31.0\,MPa$, $E = 32.25\,GPa$[†])							
1	500	100	100	0.500	⋆	1.018	⋆
2	2000	200	400	0.500	⋆	1.341	⋆
3	4000	400	800	0.500	⋆	1.278	⋆
4	2000	400	800	0.500	⋆	1.205	⋆
(Mortar, $g = 2mm, w/c = 0.54$, $f'_c = 35.0\,MPa$, $E = 25.65\,GPa$[‡])							
5	500	100	100	0.500	⋆	0.645	⋆
6	2000	200	400	0.500	⋆	0.787	⋆
7	4000	400	800	0.500	⋆	0.851	⋆
8	2000	400	800	0.500	⋆	⋆	⋆

Notes: [†] Total number of specimens tested = 17
[‡] Total number of specimens tested = 11
⋆ Values not available

APPENDIX II

This appendix gives details of test data (with source names) used in the analysis according to ECM and, whenever possible for the comparison with TPM (Table 4).

Table 22. (Nallathambi (1986), $g = 2mm, w/c = 0.5, f_c' = 44.3MPa$, $E = 26.3GPa$)

Serial	Beam dimensions, mm			a_0/W	a_e/W	K_{Ic}^e	$\bar{K}_{Ic}^e$
No.	S	B	W			$(MPa\sqrt{m})$	
1	200	40	51	0.200	0.267	0.596	0.666
2	200	40	51	0.300	0.358	0.577	0.654
3	200	40	51	0.400	0.444	0.567	0.647
4	200	40	51	0.500	0.532	0.544	0.627
5	200	40	51	0.600	0.626	0.524	0.615
6	400	55	64	0.200	0.267	0.677	0.761
7	400	55	64	0.300	0.358	0.645	0.736
8	400	55	64	0.400	0.445	0.623	0.718
9	400	55	64	0.500	0.535	0.593	0.691
10	400	55	64	0.600	0.632	0.613	0.730
11	600	80	76	0.200	0.267	0.738	0.827
12	600	80	76	0.300	0.357	0.721	0.819
13	600	80	76	0.400	0.442	0.718	0.824
14	600	80	76	0.500	0.529	0.692	0.804
15	600	80	76	0.600	0.622	0.665	0.791

Notes:
1. For each a_0/W six beams were tested.
2. Each entry in the columns K_{Ic}^e and $\bar{K}_{Ic}^e$ is actually the average of six values.

Table 23. (Nallathambi (1986), $g = 5mm, w/c = 0.5, f'_c = 42.1MPa$, $E = 29.3GPa$)

Serial	Beam dimensions, mm			a_0/W	a_e/W	K^e_{Ic}	$\bar{K}^e_{Ic}$
No.	S	B	W			\multicolumn{2}{c}{$(MPa\sqrt{m})$}
1	200	40	51	0.200	0.275	0.602	0.674
2	200	40	51	0.300	0.369	0.582	0.661
3	200	40	51	0.400	0.458	0.571	0.653
4	200	40	51	0.500	0.552	0.541	0.626
5	400	55	64	0.200	0.274	0.663	0.747
6	400	55	64	0.300	0.366	0.666	0.761
7	400	55	64	0.400	0.457	0.624	0.720
8	400	55	64	0.500	0.547	0.618	0.722
9	400	55	64	0.600	0.641	0.617	0.736
10	600	80	76	0.200	0.272	0.759	0.851
11	600	80	76	0.300	0.366	0.712	0.810
12	600	80	76	0.400	0.455	0.703	0.807
13	600	80	76	0.500	0.547	0.666	0.777
14	600	80	76	0.600	0.643	0.656	0.784

Notes:

1. For each a_0/W six beams were tested.
2. Each entry in the columns K^e_{Ic} and $\bar{K}^e_{Ic}$ is actually the average of six values.

Table 24. (Nallathambi (1986), $g = 10mm, w/c = 0.5, f'_c = 40.3MPa$, $E = 32.0GPa$)

Serial	Beam dimensions, mm			a_0/W	a_e/W	K^e_{Ic}	$\bar{K}^e_{Ic}$
No.	S	B	W			\multicolumn{2}{c}{$(MPa\sqrt{m})$}
1	200	40	64	0.200	0.281	0.623	0.690
2	200	40	76	0.200	0.280	0.678	0.742
3	200	40	76	0.300	0.376	0.649	0.720
4	200	40	102	0.200	0.308	0.790	0.850
5	200	40	102	0.300	0.376	0.710	0.771
6	200	40	102	0.400	0.465	0.729	0.796
7	200	40	102	0.500	0.558	0.718	0.792
8	400	55	64	0.200	0.281	0.682	0.769
9	600	80	76	0.200	0.280	0.742	0.833
10	600	80	76	0.300	0.375	0.735	0.836

Notes:

1. For each a_0/W six beams were tested.
2. Each entry in the columns K^e_{Ic} and $\bar{K}^e_{Ic}$ is actually the average of six values.

Table 25. (Nallathambi (1986), $g = 14mm, w/c = 0.5, f'_c = 40.9MPa,$
$E = 32.6GPa$)

Serial	Beam dimensions, mm			a_0/W	a_e/W	K^e_{Ic}	$\bar{K}^e_{Ic}$
No.	S	B	W			$(MPa\sqrt{m})$	
1	400	55	102	0.200	0.284	0.718	0.805
2	400	55	102	0.300	0.382	0.698	0.794
3	400	55	127	0.200	0.286	0.735	0.815
4	400	55	127	0.300	0.381	0.751	0.844
5	400	55	127	0.400	0.474	0.736	0.833

Notes:

1. For each a_0/W six beams were tested.
2. Each entry in the columns K^e_{Ic} and $\bar{K}^e_{Ic}$ is actually the average of six values.

Table 26. (Nallathambi (1986), $g = 20mm$, Series I, $w/c = 0.5, f'_c = 37.6MPa,$
$E = 33.0GPa$)

Serial	Beam dimensions, mm			a_0/W	a_e/W	K^e_{Ic}	$\bar{K}^e_{Ic}$
No.	S	B	W			$(MPa\sqrt{m})$	
1	600	80	127	0.200	0.292	0.733	0.828
2	600	80	127	0.300	0.389	0.761	0.870
3	600	80	153	0.200	0.290	0.809	0.908
4	600	80	153	0.300	0.388	0.801	0.911
5	600	80	153	0.400	0.485	0.775	0.889

Notes:

1. For each a_0/W six beams were tested.
2. Each entry in the columns K^e_{Ic} and $\bar{K}^e_{Ic}$ is actually the average of six values.

Table 27. (Nallathambi (1986), $g = 20mm$, Series II, $w/c = 0.5$, $f_c' = 38.0MPa$, $E = 33.2GPa$)

Serial	Beam dimensions, mm			a_0/W	a_e/W	K_{Ic}^e	$\bar{K}_{Ic}^e$
No.	S	B	W			$(MPa\sqrt{m})$	
1	1000	80	140	0.200	0.293	0.743	0.838
2	1000	80	140	0.300	0.392	0.747	0.853
3	1200	80	200	0.200	0.291	0.868	0.981
4	1200	80	200	0.300	0.389	0.882	1.011
5	1200	80	200	0.400	0.484	0.865	1.000
6	1200	80	200	0.500	0.579	0.887	1.042
7	1500	80	240	0.200	0.293	0.886	1.001
8	1500	80	240	0.300	0.390	0.912	1.045
9	1500	80	240	0.400	0.489	0.863	0.999
10	1500	80	240	0.500	0.589	0.850	1.001
11	1500	80	240	0.600	0.696	0.859	1.044
12	1800	80	300	0.200	0.295	0.904	1.023
13	1800	80	300	0.300	0.394	0.918	1.052
14	1800	80	300	0.400	0.490	0.922	1.067
15	1800	80	300	0.500	0.589	0.921	1.085
16	1800	80	300	0.600	0.691	0.966	1.171
17	1500	80	300	0.300	0.394	0.923	1.057
18	1200	80	300	0.300	0.392	0.945	1.076
19	1000	80	300	0.300	0.392	0.943	1.064

Notes:

1. For each a_0/W six beams were tested.

2. Each entry in the columns K_{Ic}^e and $\bar{K}_{Ic}^e$ is actually the average of six values.

Table 28. (Nallathambi (1986), $g = 20mm$)

Serial	Beam dimensions, mm			a_0/W	a_e/W	K_{Ic}^e	$\bar{K}_{Ic}^e$
No.	S	B	W			$(MPa\sqrt{m})$	
(RRG, $w/c = 0.50, f_c' = 41.3MPa, E = 33.0GPa$)							
1	600	80	102	0.200	0.290	0.742	0.839
2	600	80	102	0.300	0.388	0.738	0.845
3	600	80	102	0.400	0.485	0.714	0.826
4	600	80	102	0.500	0.588	0.669	0.788
5	600	80	102	0.600	0.687	0.716	0.867
(RRG, $w/c = 0.55, f_c' = 35.8MPa, E = 28.2GPa$)							
6	600	80	102	0.200	0.295	0.653	0.739
7	600	80	102	0.300	0.390	0.718	0.823
8	600	80	102	0.400	0.491	0.651	0.753
9	600	80	102	0.500	0.595	0.619	0.730
10	600	80	102	0.600	0.690	0.700	0.848
(RRG, $w/c = 0.60, f_c' = 30.2MPa, E = 23.6GPa$)							
11	600	80	102	0.200	0.298	0.598	0.678
12	600	80	102	0.300	0.396	0.635	0.728
13	600	80	102	0.400	0.495	0.616	0.713
14	600	80	102	0.500	0.606	0.555	0.656
15	600	80	102	0.600	0.703	0.633	0.771
(RRG, $w/c = 0.65, f_c' = 24.9MPa, E = 21.0GPa$)							
16	600	80	102	0.200	0.302	0.544	0.616
17	600	80	102	0.300	0.406	0.535	0.614
18	600	80	102	0.400	0.500	0.576	0.668
19	600	80	102	0.500	0.610	0.535	0.634
20	600	80	102	0.600	0.712	0.590	0.720

Notes:

1. RRG = Rounded River Gravel.

2. For each a_0/W four beams were tested.

3. Each entry in the columns K_{Ic}^e and $\bar{K}_{Ic}^e$ is actually the average of four values.

4. K_{Ic}^e (or $\bar{K}_{Ic}^e$) given in this table are not valid because the size of the specimen is outside the range of effective crack model. These values are therefore not included in the summary Table 4, but are reproduced here to give some indication of the variation of fracture toughness with w/c ratio.

Table 29. (Nallathambi (1986), $g = 20mm$)

Serial	Beam dimensions, mm			a_0/W	a_e/W	K_{Ic}^e	$\bar{K}_{Ic}^e$
No.	S	B	W			($MPa\sqrt{m}$)	
	(CRG, $w/c = 0.50, f_c' = 38.5MPa, E = 37.4GPa$)						
1	600	80	102	0.200	0.285	0.836	0.945
2	600	80	102	0.300	0.380	0.866	0.991
3	600	80	102	0.400	0.472	0.852	0.985
4	600	80	102	0.500	0.572	0.797	0.935
5	600	80	102	0.600	0.679	0.764	0.923
	(CRG, $w/c = 0.55, f_c' = 34.2MPa, E = 32.6GPa$)						
6	600	80	102	0.200	0.289	0.751	0.849
7	600	80	102	0.300	0.383	0.816	0.934
8	600	80	102	0.400	0.479	0.772	0.892
9	600	80	102	0.500	0.581	0.719	0.844
10	600	80	102	0.600	0.684	0.732	0.886
	(CRG, $w/c = 0.60, f_c' = 28.0MPa, E = 28.5GPa$)						
11	600	80	102	0.200	0.292	0.707	0.800
12	600	80	102	0.300	0.388	0.738	0.845
13	600	80	102	0.400	0.489	0.670	0.776
14	600	80	102	0.500	0.585	0.694	0.816
15	600	80	102	0.600	0.692	0.687	0.834
	(CRG, $w/c = 0.65, f_c' = 23.8MPa, E = 24.0GPa$)						
16	600	80	102	0.200	0.301	0.560	0.634
17	600	80	102	0.300	0.396	0.635	0.728
18	600	80	102	0.400	0.497	0.601	0.696
19	600	80	102	0.500	0.602	0.581	0.686
20	600	80	102	0.600	0.705	0.624	0.761

Notes:

1. CRG = Crushed River Gravel.

2. For each a_0/W four beams were tested.

3. Each entry in the columns K_{Ic}^e and $\bar{K}_{Ic}^e$ is actually the average of four values.

4. K_{Ic}^e (or $\bar{K}_{Ic}^e$) given in this table are not valid because the size of the specimen is outside the range of effective crack model. These values are therefore not included in the summary Table 4, but are reproduced here to give some indication of the variation of fracture toughness with w/c ratio.

Table 30. (Refai and Swartz (1987) $g = 19mm$, Notched Beams)

Serial No.	Beam dimensions, mm			a_0/W	a_e/W	$\underline{a}/W$	K^e_{Ic}	$\bar{K}^e_{Ic}$	K^s_{Ic}
	S	B	W					($MPa\sqrt{m}$)	
(Mix A. $w/c = 0.5$, $f'_c = 55.8\,MPa$, $E = 36.8\,GPa$)									
1*	381	76	102	0.290	0.382	0.550	0.743	0.843	1.153
2*	381	76	102	0.320	0.407	0.560	0.796	0.904	1.184
3*	381	76	102	0.460	0.541	0.680	0.787	0.906	1.158
4*	381	76	102	0.520	0.602	0.690	0.773	0.900	0.939
5*	381	76	102	0.620	0.720	0.720	0.724	0.872	0.594
6*	381	76	102	0.670	0.805	0.810	0.628	0.778	0.560
(Mix B. $w/c = 0.5$, $f'_c = 53.1\,MPa$, $E = 38.4\,GPa$)									
7	762	76	203	0.290	0.360	0.410	0.968	1.095	1.054
8	762	76	203	0.300	0.366	-	1.019	1.155	-
9	762	76	203	0.500	0.552	-	0.909	1.048	-
(Mix C. $w/c = 0.5$, $f'_c = 54.4\,MPa$, $E = 39.3\,GPa$)									
10	1143	76	305	0.300	0.369	0.420	1.275	1.444	1.378
11	1143	76	305	0.500	0.550	0.630	1.120	1.291	1.250
12	1143	76	305	0.670	0.729	0.730	1.039	1.255	0.755

Notes:

1. K^s_{Ic} values are as given by Refai and Swartz (1987).

2. * Specimen size outside the valid range for effective crack model.

Table 31. (Refai and Swartz (1987), $g = 19mm$, Mix B. Precracked Beams $f'_c = 53.1 MPa, E = 38.4 GPa$)

Serial No.	Beam dimensions, mm			a_0/W	a_e/W	a/W	K^e_{Ic}	K^e_{Ic}	K^s_{Ic}
	S	B	W				$(MPa\sqrt{m})$		
1	762	76	203	0.150	0.231	-	0.823	0.910	-
2	762	76	203	0.156	0.238	-	0.800	0.887	-
3	762	76	203	0.188	0.270	-	0.816	0.911	-
4	762	76	203	0.206	0.284	-	0.916	1.026	-
5	762	76	203	0.231	0.316	-	0.748	0.842	-
6	762	76	203	0.246	0.331	-	0.753	0.849	-
7	762	76	203	0.250	0.331	0.440	0.897	1.012	1.142
8	762	76	203	0.257	0.337	-	0.822	0.928	-
9	762	76	203	0.270	0.352	-	0.772	0.873	-
10	762	76	203	0.275	0.352	0.480	0.863	0.977	1.160
11	762	76	203	0.282	0.355	-	0.921	1.042	-
12	762	76	203	0.293	0.386	0.620	0.608	0.690	1.124
13	762	76	203	0.334	0.409	-	0.818	0.930	-
14	762	76	203	0.375	0.448	-	0.794	0.905	-
15	762	76	203	0.375	0.448	-	0.794	0.905	-
16	762	76	203	0.425	0.498	-	0.747	0.856	-
17	762	76	203	0.443	0.517	-	0.725	0.832	-
18	762	76	203	0.454	0.532	0.720	0.685	0.788	1.156
19	762	76	203	0.475	0.533	-	0.867	0.997	-
20	762	76	203	0.500	0.566	-	0.777	0.898	-
21*	762	76	203	0.563	0.628	0.780	0.773	0.905	1.212
22*	762	76	203	0.608	0.678	0.830	0.755	0.897	1.264

Notes:

1. Initial crack of depth a_0 was created by load recycling the specimen. There was no noticeable difference in K^s_{Ic} values between notched and precracked beams (cf. Mix B. Table 30).

2. K^s_{Ic} values are as given by Refai and Swartz (1987).

3. * Specimen size outside the valid range for effective crack model.

Table 32. (Refai and Swartz (1987), $g = 19mm$, Mix C. Precracked Beams $f'_c = 54.4MPa, E = 39.3GPa$)

Serial	Beam dimensions, mm			a_0/W	a_e/W	$\underline{a}/W$	K^e_{Ic}	$\bar{K}^e_{Ic}$	K^s_{Ic}
No.	S	B	W				$(MPa\sqrt{m})$		
1	1143	76	305	0.187	0.269	0.400	0.995	1.110	1.319
2	1143	76	305	0.208	0.297	0.550	0.830	0.931	1.228
3	1143	76	305	0.221	0.305	0.430	0.939	1.055	1.224
4	1143	76	305	0.250	0.333	0.490	0.937	1.057	1.337
5	1143	76	305	0.258	0.339	0.460	0.963	1.088	1.243
6	1143	76	305	0.266	0.347	0.420	0.965	1.091	1.094
7	1143	76	305	0.276	0.356	-	0.967	1.094	-
8	1143	76	305	0.313	0.393	-	0.928	1.054	-
9	1143	76	305	0.321	0.395	-	1.027	1.166	-
10	1143	76	305	0.362	0.437	-	0.946	1.078	-
11	1143	76	305	0.396	0.466	-	0.974	1.113	-
12	1143	76	305	0.411	0.474	0.580	1.057	1.208	1.309
13	1143	76	305	0.416	0.478	-	1.062	1.214	-
14	1143	76	305	0.437	0.523	-	0.752	0.864	-
15	1143	76	305	0.454	0.508	0.600	1.115	1.278	1.312
16	1143	76	305	0.479	0.540	0.620	0.999	1.150	1.138
17	1143	76	305	0.500	0.565	0.650	0.945	1.092	1.052
18	1143	76	305	0.507	0.563	-	1.034	1.195	-
19	1143	76	305	0.545	0.606	0.710	0.968	1.128	1.138
20	1143	76	305	0.570	0.625	0.690	1.024	1.198	1.033

Table 33. (Malwar (1987), $g = 10mm$)

Serial No.	Beam dimensions, mm			a_0/W	a_e/W	K_{Ic}^e	K_{Ic}^e
	S	B	W			$(MPa\sqrt{m})$	
Series I $f_c' = 29.0 MPa, E = 21.7 GPa, w/c = 0.60$							
1	788	102	102	0.500	0.545	0.764	0.890
2	788	102	102	0.500	0.539	0.816	0.949
3	788	102	102	0.500	0.543	0.780	0.909
4	788	102	102	0.500	0.547	0.741	0.864
5	788	102	102	0.500	0.543	0.780	0.909
6	788	102	102	0.500	0.549	0.724	0.844
7	788	102	102	0.500	0.540	0.809	0.942
8	788	102	102	0.500	0.543	0.778	0.906
9	788	102	102	0.500	0.551	0.704	0.822
10	788	102	102	0.500	0.545	0.757	0.882
11	788	102	102	0.500	0.551	0.682	0.796
12	788	102	102	0.500	0.543	0.780	0.909
Series II $f_c' = 58.9 MPa, E = 24.5 GPa, w/c = 0.40$							
13	788	102	102	0.300	0.374	0.804	0.916
14	788	102	102	0.300	0.365	0.983	1.118
15	788	102	102	0.300	0.367	0.910	1.035
16	788	102	102	0.300	0.367	1.020	1.161
17	788	102	102	0.500	0.539	0.825	0.960
18	788	102	102	0.500	0.544	0.879	1.024
19	788	102	102	0.500	0.549	0.931	1.086
20	788	102	102	0.500	0.539	0.911	1.061
Series III $f_c' = 33.1 MPa, E = 19.7 GPa, w/c = 0.55$							
21	788	102	216	0.500	0.543	0.967	1.112
22	788	102	216	0.500	0.544	0.962	1.107
23	788	102	216	0.500	0.549	0.908	1.045
24	788	102	216	0.500	0.545	0.947	1.090
25	788	102	102	0.500	0.540	0.808	0.940
26	788	102	102	0.500	0.543	0.750	0.874
27	788	102	102	0.500	0.540	0.778	0.906
28	788	102	102	0.500	0.543	0.749	0.873

Table 34. (Ferrara (1987))

Serial	Beam dimensions, mm			a_0/W	a_e/W	K_{Ic}^e	K_{Ic}^e
No.	S	B	W			$(MPa\sqrt{m})$	
$g = 10mm, f_c' = 55.5MPa, E = 29.8GPa, w/c = 0.48$							
1	800	100	100	0.500	0.548	0.925	1.079
2	800	100	100	0.500	0.548	0.934	1.089
3	800	100	100	0.500	0.544	0.968	1.128
4	800	100	100	0.400	0.461	0.983	1.131
5	800	100	100	0.400	0.452	1.028	1.181
6	800	100	100	0.400	0.456	0.976	1.122
7	800	100	100	0.400	0.454	1.006	1.155
8	800	100	100	0.300	0.363	1.082	1.230
9	800	100	100	0.300	0.364	1.065	1.211
10	800	100	100	0.300	0.364	1.057	1.202
11	800	100	100	0.300	0.363	1.089	1.237
12	800	100	100	0.200	0.273	1.056	1.184
13	800	100	100	0.200	0.272	1.075	1.204
14	800	100	100	0.200	0.272	1.062	1.191
15	800	100	100	0.200	0.273	1.033	1.158

Note: Beams with dimensions outside the range for effective crack model
are not shown in the table.

Table 35. (Elices et al. (1987); Planas and Elices (1986))

Serial	Beam dimensions, mm			a_0/W	a_e/W	K_{Ic}^e	$\bar{K}_{Ic}^e$
No.	S	B	W			$(MPa\sqrt{m})$	
Series BN–1, $g = 20mm, f_c' = 36.2MPa, E = 23.95GPa, w/c = 0.58$							
1	1131	100	200	0.500	0.533	0.974	1.135
2	1131	100	200	0.500	0.536	0.945	1.100
3	1131	100	200	0.500	0.543	0.869	1.014
4	1131	100	200	0.500	0.539	0.915	1.066
5	1131	100	200	0.500	0.531	1.099	1.280
6	1131	100	200	0.500	0.541	0.884	1.031
7	1131	100	200	0.500	0.551	0.793	0.926
8	1386	100	300	0.500	0.541	1.185	1.377
9	1386	100	300	0.500	0.550	1.064	1.238
10	1386	100	300	0.500	0.552	1.043	1.214
11	1386	100	300	0.500	0.541	1.185	1.377
12	1386	100	300	0.500	0.542	1.175	1.365
13	1386	100	300	0.500	0.545	1.135	1.319
14	1386	100	300	0.500	0.543	1.165	1.354
Series BN–2, $g = 16mm, f_c' = 38.3MPa, E = 34.1GPa, w/c = 0.58$							
15	400	100	100	0.330	0.406	0.854[†]	0.981[†]
16	760	100	190	0.330	0.402	1.121[†]	1.286[†]
17	1440	100	360	0.330	0.404	1.386[†]	1.591[†]

Notes:

1. [†] Calculated from mean value of P_{max} for five specimens.

2. Beams with dimensions outside the range for effective crack model are not shown in the table.

Table 36. (Alexander (1987), $g = 19mm$)

Serial	Beam dimensions, mm			a_0/W	a_e/W	K_{Ic}^e	$\bar{K}_{Ic}^e$
No.	S	B	W			($MPa\sqrt{m}$)	
Series F1, $f_c' = 29.0MPa, E = 32.5GPa, w/c = 0.53$							
1	2000	100	500	0.200	0.281	1.688	1.893
2	2000	100	500	0.200	0.285	1.487	1.670
3	2000	100	500	0.400	0.475	1.463	1.678
4	2000	100	500	0.400	0.467	1.638	1.877
5	1200	100	300	0.200	0.275	1.527	1.710
6	1200	100	300	0.200	0.273	1.585	1.775
7	800	100	200	0.400	0.474	0.982	1.125
8	800	100	200	0.400	0.477	0.935	1.072
Series F2, $f_c' = 34.3MPa, E = 33.2GPa, w/c = 0.53$							
9	3200	100	800	0.200	0.280	2.013	2.258
10	3200	100	800	0.400	0.484	1.529	1.754
11	3200	100	800	0.400	0.474	1.763	2.021
12	3200	100	800	0.400	0.475	1.729	1.982
Series F3, $f_c' = 26.3MPa, E = 32.0GPa, w/c = 0.53$							
13	2000	100	500	0.200	0.284	1.417	1.591
14	2000	100	500	0.200	0.279	1.622	1.819
15	2000	100	500	0.400	0.474	1.368	1.568
16	2000	100	500	0.400	0.472	1.405	1.611
17	1200	100	300	0.200	0.284	1.147	1.287
18	1200	100	300	0.200	0.285	1.127	1.266
19	1200	100	300	0.250	0.333	1.141	1.291
20	1200	100	300	0.250	0.333	1.141	1.291
21	1200	100	300	0.250	0.329	1.252	1.416
22	1200	100	300	0.400	0.465	1.261	1.444
23	1200	100	300	0.400	0.474	1.098	1.258
24	1200	100	300	0.400	0.475	1.081	1.240
25	800	100	200	0.200	0.276	1.230	1.379
26	800	100	200	0.200	0.278	1.172	1.314
27	800	100	200	0.400	0.466	1.073	1.229
28	800	100	200	0.400	0.460	1.173	1.343

Note:
Beams with dimensions outside the range for effective crack model are not shown in the table.

Table 37. (Bascoul (1987), $g = 3.15mm$)

Serial	Beam dimensions, mm			a_0/W	a_e/W	$\underline{a}/W$	K_{Ic}^e	$\bar{K}_{Ic}^e$	K_{Ic}^s
No.	S	B	W					$(MPa\sqrt{m})$	
(Crushed Marble, $w/c = 0.53$, $f_c' = -$, $E = 36.8\,GPa$[†])									
1	160	24	40	0.250	0.305	0.323	0.938	1.057	0.982
2	160	24	39	0.263	0.320	0.307	0.885	0.999	0.854
3	160	24	40	0.274	0.324	0.301	0.840	0.949	0.792
4	320	50	80	0.512	0.562	0.550	0.970	1.124	0.931
5	320	50	80	0.500	0.544	0.545	0.910	1.051	0.914
6	320	50	80	0.512	0.555	0.562	0.900	1.042	0.923
7	320	49	80	0.262	0.328	0.310	1.074	1.214	1.025
8	320	50	83	0.263	0.325	0.308	0.940	1.061	0.899
9	320	50	83	0.264	0.327	0.310	0.908	1.024	0.869
10	320	50	80	0.125	0.182	0.154	0.850	0.931	0.779
11	320	50	83	0.156	0.217	0.191	0.905	0.999	0.843
12	320	50	83	0.144	0.204	0.176	0.987	1.086	0.911

Notes:

1. [†] Estimated from load - deflection plots.

2. Both K_{Ic}^e and K_{Ic}^s values were calculated from the $P - \delta$ and $P - CMOD$ plots.

3. An excellent agreement between K_{Ic}^e and K_{Ic}^s is evident.

Table 38. (Catalano and Ingraffea (1982))

Serial	Beam dimensions, mm			a_0/W	a_e/W	$\underline{a}/W$	K_{Ic}^e	$\bar{K}_{Ic}^e$	K_{Ic}^s
No.	S	B	W					$(MPa\sqrt{m})$	
(Mortar, $g = 6mm$, $w/c = 0.44$, $f_c' = 60.7\,MPa$, $E = 33.5\,GPa$)									
1	1321	102	305	0.197	0.292	0.278	1.281	1.445	1.236
2	1321	102	305	0.211	0.288	0.248	1.160	1.307	1.046
(Concrete, $g = 12.7mm$, $1:2:1.62$, $w/c = 0.52$, $f_c' = 45.5\,MPa$, $E = 31.0\,GPa$)									
3	1321	152	305	0.205	0.285	0.261	1.367	1.539	1.286
4	1321	152	305	0.198	0.281	0.352	1.445	1.625	1.736
5	1321	152	305	0.202	0.271	0.252	1.476	1.658	1.403
(Concrete, $g = 12.7mm$, $1:2:2.44$, $w/c = 0.52$, $f_c' = 43.4\,MPa$, $E = 31.0\,GPa$)									
6	1321	152	305	0.211	0.282	0.267	1.589	1.788	1.529
7	1321	152	305	0.202	0.279	0.244	1.647	1.852	1.504
8	1321	152	305	0.205	0.283	0.275	1.594	1.794	1.558

Notes:

1. Both K_{Ic}^e (or $\bar{K}_{Ic}^e$) and K_{Ic}^s values were calculated from the $P - \delta$ and $P - CMOD$ plots.

2. The agreement between K_{Ic}^e and K_{Ic}^s is good.

Table 39. (Jenq and Shah (1984); John and Shah (1987))

Serial	Beam dimensions, mm			a_0/W	a_e/W	$\underline{a}/W$	K_{Ic}^e	K_{Ic}^e	K_{Ic}^s
No.	S	B	W					$(MPa\sqrt{m})$	
(Concrete, $g = 19mm, w/c = 0.65, f_c' = 25.16\,MPa, E = 27.24\,GPa$)									
1	914	86	229	0.333	0.391	0.347	1.241	1.413	1.106
2	914	86	229	0.333	0.402	0.385	1.008	1.148	0.940
3	609	57	152	0.318	0.388	0.377	0.889	1.013	0.787
4	609	57	152	0.324	0.400	0.511	0.789	0.899	1.001
5*	305	29	31	0.293	0.365	0.498	0.774	0.879	0.968
6*	305	29	31	0.293	0.373	0.548	0.656	0.746	0.891
7*	305	29	31	0.293	0.369	0.535	0.721	0.819	0.989
8*	305	29	31	0.293	0.364	0.531	0.794	0.901	1.127
(High Strength Concrete, $g = 8mm, w/c = 0.22, f_c' = 110MPa, E = 56.55GPa$)									
9	203	25	76	0.329	0.375	†	1.960	2.229	2.130
10	203	25	76	0.329	0.378	†	1.833	2.086	2.130

Notes:

1. † Values not available.

2. a_e/W were calculated from the maximum loads and the E values given by Jenq and Shah (1984) and John and Shah (1987).

3. * Specimen size outside the valid range for effective crack model. However, the agreement between the predicted values of K_{Ic}^e and K_{Ic}^s for other beams is good.

4. Although the regression formula established according to the effective crack model is based on the results of normal concrete, the prediction for the high strength concrete is very good (see Sl. Nos. 9 and 10).

Table 40. (Mindess (1984), $g = 13mm, f_c' = 48.5MPa$, $E = 33.3GPa, w/c = 0.38$)

Serial	Beam dimensions, mm			a_0/W	a_e/W	K_{Ic}^e	K_{Ic}^e
No.	S	B	W				$(MPa\sqrt{m})$
1	1600	200	200	0.508	0.678	0.360	0.429
2	1600	200	200	0.512	0.624	0.602	0.706
3	1600	200	200	0.512	0.681	0.365	0.436
4	3200	400	400	0.516	0.706	0.419	0.504
5	3200	400	400	0.511	0.733	0.327	0.397
6	3200	400	400	0.509	0.686	0.455	0.543

Note: Beams with dimensions outside the range for effective crack model are not shown in the table.

Table 41. (Horvath and Persson (1984))

Serial	Beam dimensions, mm			a_0/W	a_e/W	K_{Ic}^e	K_{Ic}^e
No.	S	B	W			$(MPa\sqrt{m})$	
CNG, $g = 8mm, f_c' = 93.0MPa, E = 32.0GPa, w/c = 0.40$							
1	800	100	100	0.500	0.533	1.016	1.181
2	800	100	100	0.500	0.536	0.979	1.139
3	800	100	100	0.500	0.529	1.058	1.229
4	800	100	100	0.500	0.527	1.094	1.270
5	800	100	100	0.500	0.538	0.949	1.104
6	800	100	100	0.500	0.529	1.064	1.236
7	1131	100	200	0.500	0.533	1.330	1.547
8	1131	100	200	0.500	0.529	1.411	1.640
9	1131	100	200	0.500	0.536	1.290	1.501
10	1386	100	300	0.500	0.539	1.496	1.741
11	1386	100	300	0.500	0.539	1.486	1.730
CNG, $g = 8mm, f_c' = 28.0MPa, E = 31.0GPa, w/c = 0.80$							
12	800	100	100	0.500	0.560	0.707	0.828
13	800	100	100	0.500	0.556	0.733	0.857
14	800	100	100	0.500	0.545	0.838	0.978
15	800	100	100	0.500	0.553	0.764	0.893
16	800	100	100	0.500	0.559	0.714	0.835
17	800	100	100	0.500	0.566	0.656	0.769
18	1131	100	200	0.500	0.563	0.897	1.051
19	1131	100	200	0.500	0.566	0.872	1.021
20	1386	100	300	0.500	0.572	0.976	1.145
21	1386	100	300	0.500	0.571	0.983	1.153
22	1386	100	300	0.500	0.570	0.994	1.165
NG, $g = 12mm, f_c' = 68.0MPa, E = 39.0GPa, w/c = 0.40$							
23	1131	100	200	0.500	0.554	1.289	1.507
24	1131	100	200	0.500	0.547	1.401	1.636
25	1131	100	200	0.500	0.559	1.226	1.435
26	1386	100	300	0.500	0.561	1.421	1.663
27	1386	100	300	0.500	0.570	1.275	1.495
28	1386	100	300	0.500	0.555	1.522	1.780
NG, $g = 12mm, f_c' = 21.0MPa, E = 26.0GPa, w/c = 0.80$							
29	1131	100	200	0.500	0.575	0.684	0.803
30	1131	100	200	0.500	0.572	0.704	0.825
31	1131	100	200	0.500	0.574	0.691	0.810
32	1386	100	300	0.500	0.583	0.739	0.869
33	1386	100	300	0.500	0.577	0.786	0.923
34	1386	100	300	0.500	0.580	0.761	0.895

1. CNG – Crushed Natural Gravel.

2. NG – Natural Gravel.

3. Beams with dimensions outside the range for effective crack model are not shown in the table.

Table 42. (Hilsdorf and Brameshuber (1987))

Serial No.	Beam dimensions, mm			a_0/W	a_e/W	$\underline{a}/W$	K_{Ic}^e	$\bar{K}_{Ic}^e$	K_{Ic}^s
	S	B	W					$(MPa\sqrt{m})$	
(Concrete, $g = 32mm, w/c = 0.54, f_c' = 31.0\,MPa, E = 32.25\,GPa$[†])									
1*	500	100	100	0.500	0.555	⋆	1.106	1.293	1.018
2	2000	200	400	0.500	0.571	⋆	1.291	1.514	1.341
3	4000	400	800	0.500	0.572	⋆	1.696	1.990	1.278
4	2000	400	800	0.500	0.568	⋆	1.767	2.070	1.205
(Mortar, $g = 2mm, w/c = 0.54, f_c' = 35.0\,MPa, E = 25.65\,GPa$[‡])									
5*	500	100	100	0.500	0.531	⋆	0.730	0.842	0.645
6	2000	200	400	0.500	0.564	⋆	0.967	1.122	0.787
7	4000	400	800	0.500	0.580	⋆	1.142	1.328	0.851
8	2000	400	800	0.500	0.575	⋆	1.196	1.390	⋆

Notes:

1. [†] Total number of specimens tested = 17

2. [‡] Total number of specimens tested = 11

3. ⋆ Values not available

4. * Specimen size outside the valid range for effective crack model.

5. The K_{Ic}^e (or $\bar{K}_{Ic}^e$) values may be inaccurate because the maximum loads were estimated in an inverse manner from the published values of K_{Ic}.

Table 43. Details of test groups used in the comparative study according to ECM and SEL (Table 5).

Group No.	g (mm)	No. of specimens tested	Size, mm S B W	Source
G-1	20	6	0600 80 076	Nallathambi(1986)
		4	1000 80 140	Nallathambi and
		4	1200 80 200	Karihaloo(1986a,b)
		4	1500 80 240	
		4	1800 80 300	
G-2	20	6	0600 80 076	
		4	1000 80 140	
		4	1200 80 200	
		4	1500 80 240	
		4	1800 80 300	
G-3	20	6	0600 80 076	
		4	1000 80 140	
		4	1200 80 200	
		4	1500 80 240	
		4	1800 80 300	
G-4	13	3	095 38 038	Bažant and
		3	191 38 076	Pfeiffer(1987)
		3	381 38 152	
		3	762 38 305	
G-5	5	3	095 38 038	
		3	191 38 076	
		3	381 38 152	
		3	762 38 305	
G-6	19	3	0400 100 100	Alexander(1987)
		2	0800 100 200	
		2	2000 100 500	
G-7	19	8	0400 100 100	
		2	0800 100 200	
		2	1200 100 300	
		1	2000 100 500	
		1	3200 100 800	

2 FRACTURE PROPERTIES OF CONCRETE AS DETERMINED BY MEANS OF WEDGE SPLITTING TESTS AND TAPERED DOUBLE CANTILEVER BEAM TESTS

P. ROSSI
Laboratoire Central des Ponts et Chaussées (LCPC), Paris, France
E. BRÜHWILER
University of Colorado, Boulder, USA (formerly: Laboratoire des
Matériaux de Construction (LMC), Ecole Polytéchnique Fédérale
de Lausanne (EPFL), Switzerland
S. CHHUY
Laboratoire Régional, Melun, France
Y.-S. JENQ
Ohio State University, Columbus, USA
S.P. SHAH
Northwestern University, Evanston, USA

1 Introduction

1.1 Objectives

Because of the testing simplicity, the Three Point Bend Test (TPBT) recommended by RILEM Technical Committee 50-FMC (RILEM 1985) is widely used for the determination of fracture properties. However, most experiments were performed on relatively small beams seldomly exceeding 200 mm in depth, and, the self-weight compared to the fracture area of TPB beams is large. The effect of self-weight on fracture properties must be carefully considered in the evaluation of fracture properties using the TPBT. Additionally, although the TPB beam can be handled in the laboratory, it does not lend itself either to the fabrication on the building site or to the use of material drilled from existing structures. These drawbacks led to the formation of Subcommittee B within RILEM TC 89-FMT with the main task to propose a testing method as an alternative to the TPBT. Simple "compact" specimen shapes such as cubes and cylinders should be used in such an alternative testing method, and the test should easily be carried out in a "normally equipped" civil and materials engineering laboratory.

The experimental work in this Subcommittee B has been performed jointly by the "Laboratoire des Matériaux de Construction (LMC)" of the Swiss Federal Institute of Technology (EPFL) in Lausanne, Switzerland, the "Laboratoire Régional" in Melun, France, and the "Laboratoire Central des Ponts et Chaussées" in Paris (LCPC), France. The objectives of this report are:

- the determination of fracture properties by means of the Wedge Splitting Test using cylindrical and cubic specimens,

- the comparison with results obtained from large Tapered Double Cantilever Beams of same concrete,

- the investigation of the influence of specimen size and maximum aggregate size on fracture properties of concrete, and,

- the comparison of different methods of evaluation of fracture properties.

1.2 Application of fracture properties

The most significant implication of brittle behaviour of concrete structures is the structural size effect (e.g. Bažant 1984) which implies that most results of concrete fracture experiments cannot be uniquely described by the commonly used strength of materials (SOM) approach. Linear elastic fracture mechanics (LEFM) captures the size effect, but overestimates the load-bearing capacity of structures of usual dimensions. Rather, there exists a gradual transition from the SOM criterion to LEFM. The transition can be modelled by nonlinear fracture mechanics (NLFM) approaches. In this transition range, the size of the fracture process zone, or the "region of discontinuous microcracking ahead of a stress free crack (Mindess 1989)" in the concrete specimen, is large compared to the structural size.

Size effects must be considered in the analysis of structures, and the application of fracture properties is strongly linked to the size of the concrete structure to be analysed and the fracture model to be used. Based on experiments and size effect considerations, the limits of applicability of classical LEFM to concrete dams were assessed (Brühwiler et al. 1989), and it was found that LEFM can be used to analyse the bottom part of large gravity dams. However, NLFM should be used for fracture mechanics analysis of arch/buttress dams and the top part of gravity dams. Consequently, all reinforced concrete structures which are smaller structures than dams, can be expected to fall in the range of NLFM models.

The most often used concept in NLFM is the concept of fracture energy with the specific fracture energy G_F and the tensile softening diagram as principal fracture properties. In order to describe the failure behaviour of concrete structures,

Hillerborg, Modéer and Petersson (1976) and Bažant and Oh (1983) developed the Fictitious Crack Model (FCM) and the Blunt Crack Model, respectively. Both models are based on the tensile strength, f_t, the specific fracture energy, G_F, and the tensile softening diagram of the material. Experimental results reveal that G_F, in general, depends on the maximum aggregate size and specimen dimension (Hordijk et al. 1989), that is, G_F grows with both increasing maximum grain size and fracture ligament length of the specimen. However, G_F-values independent of specimen dimension (fracture ligament length) were obtained from specimens with ligament lengths longer than 300 mm (Wittmann et al. 1988, Saouma et al. 1989). Additionally, as in some investigations (e.g. Saouma et al. 1989) no influence of maximum aggregate size on G_F was observed, the size as well as the shape and quality of aggregates seem to affect fracture properties of concrete. In addition to the value of G_F, the tensile softening diagram is needed for application of the concept of fracture energy in computerized structural analysis. The most direct way to determine the tensile softening diagram of concrete is the uniaxial tensile test which is difficult to carry out. To overcome this difficulty, simplified models representing the real softening behaviour are used. A finite element analysis allowing the determination of a bilinear softening diagram from the experimentally obtained load-displacement curves by means of a data fit, is presented in this report.

Another NLFM approach is the Two Parameter Fracture Model (Jenq and Shah 1985) which assumes a modified Griffith crack, (i.e., a traction-free crack). Two fracture parameters, critical stress intensity factor, K_{Ic}^s, and critical crack tip opening displacement, $CTOD_c$, are needed in this model. Although K_{Ic}^s and $CTOD_c$ have been shown to be independent of specimen size using TPBT, the sensitivity of the fracture parameters to different testing geometries is yet to be determined.

The critical stress intensity factor, or fracture toughness K_{Ic}, is used in a classical LEFM structural analysis. For the determination of K_{Ic}, LEFM can be applied to concrete specimens only if the dimensions of the specimen used for the determination of the fracture toughness of the concrete are much larger than the largest aggregate in the concrete (Entov and Yagust 1975, Rossi 1986, Rossi et al. 1986). This condition implies very large specimens, and, modified LEFM in which an effective crack is considered in order to obtain an "objective" fracture toughness K_{Ic}. For example, Rossi (1986) obtained an objective fracture toughness K_{Ic} using the Compliance Method for common structural concrete with a Double Cantilever Beam (DCB) specimens 3.5 m long and 1.1 m wide. This test requires large resources and would be difficult to carry out in a normally equipped civil engineering laboratory. The logical approach from the experimental viewpoint is to carry out parallel testing programmes in which small specimens yielding "non-objective" K_{Ic}-values where the scale effect appears, and tests on large specimens yielding "objective" K_{Ic}-values where the scale effect is eliminated. Then an analytical relation between the values of K_{Ic}, as determined by these two testing programmes may be established. If this approach is successful, the objective fracture toughness K_{Ic} of in-situ concrete can be determined on the basis of tests on extracted cores.

1.3 Outline of report

In this report, Wedge Splitting Tests, performed by E. Brühwiler at LMC-EPFL, and Tapered Double Cantilever Beam Tests, carried out by S. Chhuy at L.R. Melun, are described in Sections 2 and 3. The G_F-values and the tensile softening diagrams were evaluated at LMC-EPFL (Section 4), and the fracture toughness K_{Ic} of all tests was determined by P. Rossi using the compliance method (Section 5). Fracture properties according to the Two Parameter Model were evaluated by Y. Jenq (Section 6), and are presented in a separate report (Jenq and Shah 1989). Finally, conclusions and recommendations for future work based on the entire investigation will be made in Section 7.

2 Description of experiments

2.1 Wedge splitting tests (E. Brühwiler)

2.1.1 Wedge splitting specimens

The Wedge Splitting specimen shape, previously proposed by Linsbauer and Tschegg (1986), is characterized by a groove and a starter notch which can be either moulded or saw cut. Sketches of the specimen shapes are shown in Figure 1. The cubical specimen A is suitable for freshly poured concrete and can be fabricated in standard moulds. The cylindrical specimens B and C are suggested to examine drilled cores from existing structures. The main drawback of specimen C is that it needs either a deep notch or a longitudinal groove on both sides in order to prevent shear failure of one of the cantilevers. Additionally, evaluation of fracture properties using FE-analysis would enhance a 3-dimensional analysis. In principle, specimens can be fabricated from concrete blocks of any shape as shown in shape D where two plane parallel surfaces are polished, and, the groove and notch are sawn.

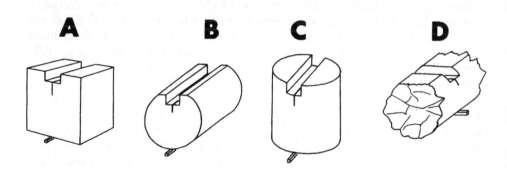

Figure 1: Wedge splitting specimen shapes.

Contrary to the frequently used three-point bending beams, WS specimens can be easily fabricated on the building site (cubes/cylinders) where they match the geometry of standardized specimens for freshly poured concrete. Also cores (cylinders) can be drilled on site which fits with the objective of taking samples from existing structures. Preparation of the specimen with groove and notch requires a saw available in a "normally equipped" laboratory. Additionally, the specimens are easy to handle, and there is no risk of breaking them during handling. The dimensions of the cubic and cylindrical WS specimens used in the present investigation are given in Figure 2.

The ratio of fracture area to specimen weight is important with regard to the investigation of size effects (large specimens) (Saouma *et al.* 1989) and the testing of concretes with large aggregates (mass concrete) (Saouma *et al.* 1989, Brühwiler 1988).The fracture area of WS specimens is large compared to the self-weight. In order to illustrate this feature, a WS cube having a side length of 200 mm is compared to the three-point bending beam with a span of 800 mm and dimensions according to the RILEM Recommendation (RILEM 1985). Both specimens have the same weight (20 kg). However, the fracture area of the WS cube with a fracture ligament of 130 mm is 5.2 times larger than the fracture area of the beam.

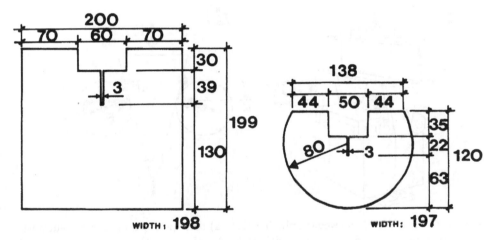

Figure 2: Geometry of the tested cubical and cylindrical WS specimen (Dimensions in mm).

2.1.2 Test method and design of the test set-up

The principle of the Wedge Splitting Test (WST) (Brühwiler and Wittmann 1990, Brühwiler 1988) is schematically presented in Figure 3. First, the WS specimen is placed on a linear support which is fixed to the lower plate of the testing machine (Fig. 3a). Two massive steel loading devices equipped with roller or needle bearings on each side are placed on the top of the specimen (Fig. 3b). A steel profile with two identical wedges is fixed at the upper plate of the testing machine. The actuator of the testing machine is moved so that the wedges enter between the bearings resulting in a horizontal splitting force component. The dimensions of groove and notch are chosen so that the crack propagates in the vertical direction, and the specimen is split into two halves. The fracture section of the specimen is essentially subjected to a bending moment (Regnault and Brühwiler 1990). The main feature of the loading device is the use of wedges and bearings. The same principle was previously used by Hillemeier and Hilsdorf (1977).

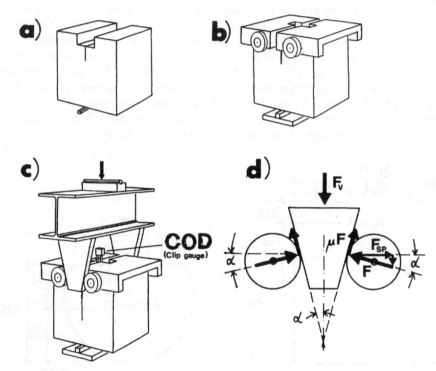

Figure 3: Principle of the wedge splitting test: a) test specimen on a linear support, b) placing of two loading devices with roller bearings, c) the wedges are pressed between the bearings in order to split the specimen into two halves, and, d) forces acting on the wedge.

During the test, the load in the vertical direction and the crack opening displacement (COD) are monitored. The splitting force, F_s, is the horizontal component of the force acting on the bearings and is calculated taking the wedge angle and frictional forces into consideration (Fig. 3d). Force equilibrium yields the splitting force:

$$F_s = \frac{F_v}{2 \cdot tan\alpha} \cdot \frac{(1 - \mu \cdot tan\alpha)}{(1 + \mu \cdot ctg\alpha)} \approx \frac{F_v}{2 \cdot tan\alpha} \cdot \frac{1}{(1 + \mu \cdot ctg\alpha)} \tag{1}$$

where α is the wedge angle according to Figure 3d, and μ is the coefficient of friction. The manufacturers of roller bearings give μ-values ranging from 0.1 % to 0.5 %. With $\alpha = 15°$, and $\mu = 0.1$ % and 0.5 %, the effect of frictional forces on F_s is about 0.4 % and 1.9 %, respectively. The friction can be reduced by (1) using hardened steel inserts along the inclined wedge surface, (2) roller or needle bearings instead of common ball bearings, and, (3) carefully polishing the wedge surface. These measures can lead to significant reduction of friction as shown by Hillemeier (1976) who experimentally determined on a calibrator a μ-value = 0.031 % for his wedge loading set-up.

Friction also occurs in the roller supports of the TPBT. For same value of μ, friction in the WST set-up is about 6-times larger than for the TPBT (Planas and Elices 1989). As frictional effects in the WST can in practice be kept down to 2 %, they can be neglected in a first approximation, and, the splitting force (evaluated in this report) is computed from the measured vertical force as:

$$F_s = \frac{F_v}{2 \cdot tan\alpha} \tag{2}$$

However, if the coefficient of friction is larger than 0.5 %, then frictional forces should be taken into consideration in the evaluation of test data.

The crack opening displacement COD is monitored using a transducer, or a clip gauge, fixed at the level where the resultant splitting force acts on the specimen. In a closed-loop servo-hydraulic testing machine, the test is performed under COD control. A "quasi-static" COD-rate of 30 μm/min was chosen in the present experiments. Stable WST can also be performed under displacement control (actuator stroke or crosshead displacement) using conventional testing machines. In this case, the interaction between testing machine stiffness, the specimen stiffness and the material properties must be considered, and the appropriate notch length of the specimen must be found to ensure a stable fracture test (Brühwiler and Wittmann 1990, Brühwiler 1988).

The performance of stable fracture mechanics tests on concrete specimens is difficult because of the high specimen stiffness compared to the testing machine stiffness and the small rupture deformations of concrete. The WST overcomes these difficulties by the use of wedges:

- In the WST only the vertical load component deforms the frame of the testing machine. Using a small wedge angle, the vertical force is reduced relative to the splitting force for a given specimen. Consequently, the test can be conducted in a machine that is not very stiff.

- The actuator displacement which is perpendicular to the specimen deformation (COD), is increased with respect to the COD if a small wedge angle is chosen.

Hence, a small wedge angle should be used to increase the stiffness of the specimen-machine assembly and to facilitate the control of the actuator displacement. However, frictional forces increase significantly with decreasing α (Eq. 1). A wedge angle of 15 ° has proved to be appropriate.

The present WST set-up was constructed using a cross-wise arrangement of the upper and lower linear supports (Fig. 3c). The advantage of this set-up consists in its statical system which is isostatic, and, hence, alignment errors are minimal. Additionally, the loading device and the beam with the two wedges are built such that adjustments necessary for different specimen widths can be easily made. The loading device with the bearings has been designed as stiff as reasonably possible in order to keep "bow-out" deformations small. These deformations are expected to be of negligible influence on the load deformation response of a specimen although there is no experimental evidence that the force is evenly distributed along the groove.

2.1.3   Evaluation of specific fracture energy

The specific fracture energy, G_F, is obtained from the area under the splitting force F_s vs. COD curve divided by the projected fracture area (ligament length · width). The "input" energy could also be determined from the area under the vertical force F_v vs. (vertical) actuator displacement, u-curve. Theoretically, both energy values should be the same, but larger energy values are determined from the F_v vs. u-curve, because u includes not only specimen deformation but also displacements in the test set-up and the testing machine. The influence of the self-weight on the fracture energy is negligible even for large specimens which is an important advantage compared to the TPBT where the fracture energy part due to the self-weight of the beam amounts to 40 - 60 % of the total fracture energy.

The vertical force component, F_v, also contributes to the fracture energy. During fracturing when the specimen opens, the load point (point where the resultant force acts on the specimen) undergoes some displacement in the vertical direction. The fracture energy in the vertical direction can be estimated from F_v times the vertical displacement which is obtained by comparing the geometrical situation of the uncracked ($COD = 0$) with the cracked (COD of some finite value) specimen. A portion of 5 % and 9 % of the fracture energy obtained under the F_s - COD curve, was estimated for the cubical and cylindrical specimen, respectively. The fracture energy in the vertical direction decreases the closer the load point is to the axis of symmetry of the specimen with respect to the specimen height (cubical specimen). The more the load point is away from the axis of symmetry, the more important is this contribution (cylindrical specimen), and, a very long WS specimen would correspond to the 3PB beam (Fig. 4). Although the fracture energy in the vertical direction is significant (according to the estimation), it is not included in the G_F-values given in this report. However, a more thorough investigation of this energy part is necessary.

For practical reasons (damaging of testing equipment when the closed loop is lost), the WS test is stopped before complete failure. For the determination of

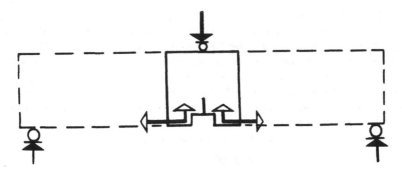

Figure 4: Wedge splitting specimen as a "compact" Three Point Bend Beam.

the total fracture energy, the end portion of the F_s - COD curve is obtained by adding the tangent to the measured curve. As this last portion of fracture energy is small compared to the total fracture energy, only a minor imprecision may be introduced. This is in contrast to the TPBT where the maximum deflection must be exactly known for the determination of the part of fracture energy due to self-weight. Measurement errors in the determination of the maximum deflection may significantly affect the evaluated fracture energy (Planas and Elices 1989).

Spurious energy dissipation may occur at the support. However in practice, this support has a finite size (e.g. a steel bar 5 mm x 5 mm), and the local stress obtained by computing $F_{v,max}$ divided by the area of the support is well in the linear elastic range. Thus, spurious energy dissipation may be of minor importance.

2.1.4 Evaluation of fracture toughness

In this report, the fracture toughness K_{Ic} for the LEFM analysis is evaluated using the compliance method (Rossi 1986) where an effective crack length which is larger than the true crack but shorter than the true crack plus the fracture process zone, is determined from finite element calibration (see Section 5). In the finite element analysis, a relationship between crack length and compliance is determined and compared to the specimen compliance which is obtained from unload/reload cycles during the test. Fracture parameters according to the Size Effect Method, the Two Parameter Fracture Model and the Effective Crack Model (see report of Subcommittee A (Karihaloo and Nallathambi 1989) in Chapter 1 of this volume) can also be evaluated if the corresponding analytical expressions for a given WS geometry are known. In Section 6.2, fracture parameters according to the Two Parameter Model are presented.

2.2 Tapered double cantilever beam tests (S. Chhuy)

The TDCB specimen shown in Figure 5 is suspended vertically. The specimen is subjected to an opening displacement (COD) using a flat jack placed in the pre-existing notch. The experiment is performed at an imposed rate of COD which is programmed by an electronic servo-control system and ensures stable crack propagation. The COD is measured by a sensor placed in the notch at the point of application of the force. In all tests, the imposed COD-rate was 15 μm/min. Loading/unloading cycles were conducted to determine the specimen compliance which varies in the course of the experiment. This information is necessary for the determination of K_{Ic} at different stages of crack propagation.

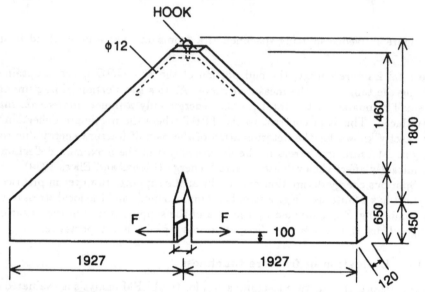

Figure 5: Geometry of the TDCB specimen.

Since the TDCB specimen is in a vertical position, the influence of the self-weight in computing of the released energy G_I is *a priori* negligible when compared to that of the force. Hand calculations show that the bending moment caused by the self-weight (at peak COD) is about 2,500 times smaller than the moment consisting of notch length times minimum peak force value obtained in the TDCB tests. Thus, the self-weight of the TDCB specimen does not significantly affect the test results.

The single reinforcing bar (diameter = 12 mm) for the fixation of the hook is located in the compression zone of the specimen. The stiffness of this reinforcing bar is about 5 orders of magnitude smaller than the initial stiffness of the specimen, and, hence, the stiffening effect of the reinforcing bar is not considered in the evaluation of test data.

2.3 Specimen preparation (S. Chhuy, P. Rossi)

2.3.1 Concrete mix

Three types of structural concretes differing in grading were examined: a "macro" (Concrete A), a "mini" concrete (Concrete B) and a "micro" concrete (Concrete C). The compositions of the three concretes are given in Table 1. These three concretes were chosen because the fracture properties of concrete (with the exception of high-strength and fibre-reinforced concrete), in general, depend on the size of their largest aggregates. A single relation between fracture properties evaluated on the basis of a large specimen and those from small specimens is found. If such a relation is valid for the three types of concrete, it may be applicable to most of the concretes in common use in civil engineering with the possible exception of mass concretes with very large aggregates.

Table 1: Concrete mix design.

	CONCRETE		
	A [kg/m^3]	B [kg/m^3]	C [kg/m^3]
CPA/HP CEMENT	400	400	400
WATER	188	190	210
WATER/CEMENT RATIO	0.47	0.48	0.53
SILICO-CALCAREOUS GRAVEL:			
0/4 mm	635	700	
4/12.5 mm	350	1108	
12.5/20 mm	810	-	
0/3.15 mm			595
3.15/6.3 mm			1188
TOTAL	2,383	2,398	2,393

2.3.2 Manufacture and conservation of the specimens

For every concrete mix, one TDCB specimen, five cubes having a side length of 200 mm and five cylinders with a diameter of 160 mm and a length of 300 mm were fabricated at L.R. Melun. After demoulding, the specimens were protected from drying with a resin coating. The cubes and cylinders were then sent to the LMC in Lausanne where they were machined to obtain the specimen geometries shown in Figure 2. The test-ready specimens were then kept in water until testing.

The "micro" concrete TDCB specimen failed during the preparation of the test, a second series of Concrete C, namely C2, was fabricated. WS Tests were carried out on both "micro" Concretes C1 and C2. Both the WS and the TDCB tests have been performed the same day. Specimens of Concrete A were tested at an age of 45 days, Concrete B at 150 days, Concrete C1 at 31 days and Concrete C2 at 48 days.

2.3.3 Material properties of the hardened concretes

The 28-days material properties of the investigated concretes was determined on cylindrical specimens with a diameter of 160 mm and a length of 320 mm. The mean values of compressive strength, f_{cc}, Young's modulus E and tensile splitting strength, $f_{t,sp}$, determined by means of the brazilian splitting test for each concrete are reported in Table 2.

Table 2: Material properties at an age of 28 days.

CONCRETE	f_{cc} [MPa]	E [MPa]	$f_{t,sp}$ [MPa]
A	51	36,600	3.7
B	55	36,000	4.5
C1	60	34,200	4.7
C2	44	32,500	3.8

3 Experimental results (P. Rossi, E. Brühwiler)

For each concrete, only one TDCB specimen was tested, but at least three tests per WS specimen geometry were carried out. This choice is justified because:

- the LCPC's experience of experiments on large specimens has indicated very low scatter, and, thus, the results are highly reproducible (Rossi 1986, Rossi et al. 1986), and,

- at least three small specimens are usually tested.

Both cylindrical and cubic wedge splitting specimens were tested using the same testing machine and the same (calibrated) clip gage to monitor the COD. Figure 6 shows the force vs. COD curves obtained from the TDCB tests of all three concretes. One representative curve of both a cubical and a cylindrical WS specimen is plotted in Figure 7. All curves show loading/unloading cycles in order to monitor changes in the compliance during crack propagation.

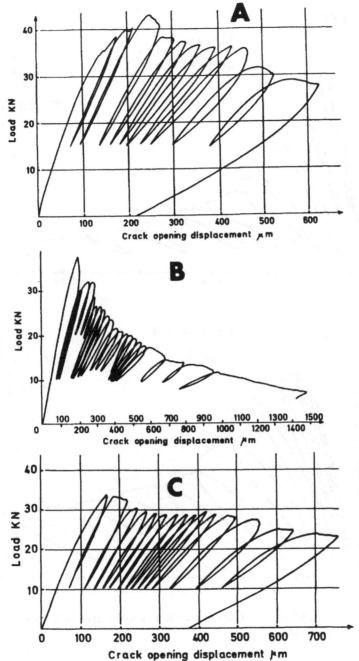

Figure 6: Force vs. COD curve of TDCB tests: Concrete A,   Concrete B, and, Concrete C2.

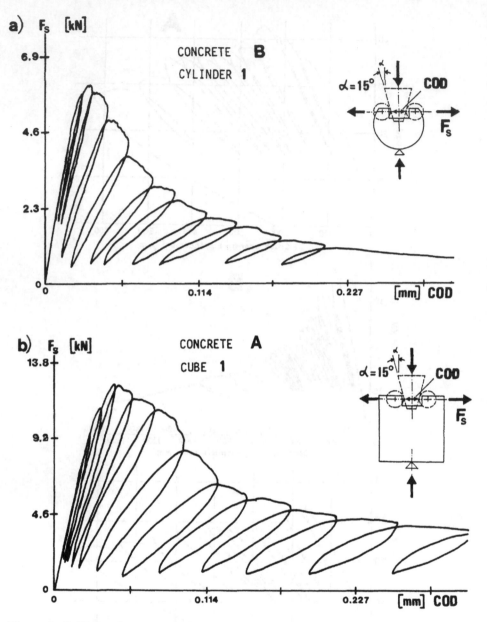

Figure 7: Splitting force vs. COD curve of a) a cylindrical WS specimen, and, b) a cubical WS specimen.

4 Determination of G_F-values and tensile softening diagrams (E. Brühwiler)

4.1 Specific fracture energy

4.1.1 Wedge splitting tests

The specific fracture energy is obtained from the area under the overall splitting force F_s-COD curves divided by the ligament area (= ligament length · specimen thickness). The single and mean values of G_F and the relative standard deviations, s, for each concrete are shown in Table 3. Mean curves were determined for both cylindrical and cubic specimens of the tested concretes (Figure 8). Each mean curve was determined from the individual $F_s - COD$ curves such that the maximum force of the mean curve corresponds to the average of the individually obtained maximum loads. Moreover, the difference between G_F-values determined from the mean curve and the average value of the G_F-values from the individual curves is smaller than 1%.

Table 3: G_F-values obtained from the Wedge Splitting Tests.

WS Specimen	Concrete	G_F [N/m]					$G_{F,mean}$ [N/m]	s [%]
Cylindrical	A	70.5	69.8	51.6	82.9	-	68.7	18.8
	B	67.9	75.6	64.1	64.4	-	68.0	7.8
	C1	51.4	67.4	58.0	62.4	-	59.8	11.4
	C2	44.0	50.1	58.7	42.4	-	48.8	15.1
Cubic	A	105.1	81.8	83.6	85.7	81.1	87.5	11.5
	B	75.0	76.2	75.1	77.1	-	75.9	1.3
	C2	72.8	81.1	64.2	-	-	72.7	11.6
	C2	51.8	45.1	67.6	-	-	54.9	21.2

The specific fracture energy G_F as a function of the maximum grain size is shown in Figure 9, and indicates that G_F increases with maximum aggregate size. This result, valid for both types of WS specimens, coincides with most investigations given in the literature. Examination of the fracture surfaces of the WS specimens revealed a large number of aggregate failures for all concretes investigated. The specimens of Concretes B and C showed a direct crack path and a "plane" fracture surface contrary to Concrete A where a more tortuous crack path was observed. The increase of G_F is explained by the more tortuous crack path when concrete with larger aggregate is tested. The difference in G_F-value of Concretes C1 and C2, having the same composition, may be explained by the poor quality of C2 due to bad compacting (Table 2).

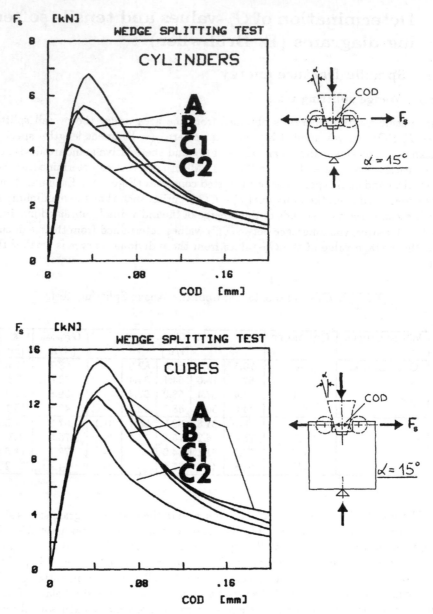

Figure 8: Splitting force vs. COD mean curves from cylindrical WS specimens and cubical WS specimens.

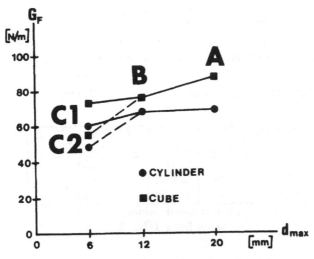

Figure 9: Specific fracture energy as function of maximum aggregate size.

Figure 9 also indicates larger G_F-values for the cubical specimen with a longer ligament length. This result, that is, G_F increases with increasing ligament length, can be explained as follows. The definition of G_F in the fictitious crack model assumes that no energy is consumed outside the fracture plane. If the fracture ligament of the specimen is increased, this assumption is not precisely fulfilled. The energy dissipation in the fracture process zone is a function of the volume, and, the stress distribution in the specimen causes energy consumption in highly stressed regions outside the fracture plane. Larger ligament lengths compared to the specimen heights result in a smaller stress gradient along the fracture ligament leading to a longer fracture process zone. The width of the fracture process zone can be assumed to increase with its length. Hence, more energy is dissipated in the fracture process zone of long fracture ligaments compared to the energy consumed in the final separation of the material. For very long ligament lengths, larger than 300 mm, G_F tends to be independent of ligament length (Wittmann *et al.* 1988, Saouma *et al.* 1989).

4.1.2 TDCB tests

A specific fracture energy value was determined from the area under the overall force - COD curve of the TDCB specimens divided by the fracture area (ligament length of 1560 mm · specimen thickness of 120 mm). Since the tests were stopped before the specimens were broken, the end part of the descending branch was completed by adding the tangent to the recorded curve. TDCB Tests on Concretes A and C2 were already stopped at a load level of about 65 % of maximum load (Fig. 6). Thus, only a relatively small portion of the descending branch was recorded, and, therefore, the experimental G_F-value for Concretes A and C2 were obtained by estimating the second portion of the descending branch. (The work of fracture from the second portion of the descending branch was 19 % and 20 % of the total fracture energy for Concretes A and B, respectively.) The G_F-values obtained from the TDCB are given in Table 4. As the TDCB is a relatively large specimen, approaching the range

Table 4: G_F-values obtained from TDCB tests.

CONCRETE	G_F [N/m]	$E^{b)}$ [MPa]	$K_{Ic}^{LEFM} = \sqrt{G_F \cdot E}$ [$MPa\sqrt{m}$]
A	$150^{a)}$	36,600	2.34
B	141	36,000	2.25
C2	$136^{a)}$	32,500	2.10

a) estimated values

b) taken from Table 2

of validity of classical LEFM, the specific fracture energy G_F can be converted to the fracture toughness using the LEFM formula (see section 5.1):

$$K_{Ic}^{LEFM} = \sqrt{G_{Ic} \cdot E} = \sqrt{G_F \cdot E} \tag{3}$$

where G_{Ic} is the critical energy restitution ratio. Table 4 shows K_{Ic}^{LEFM}-values for the TDCB specimens obtained using G_F and Eq. 3. For Concretes A and B, K_{Ic}^{LEFM} is 11 % and 2 %, respectively, larger than K_{Ic} evaluated using the compliance method (see section 5.3.2). This small difference between the two fracture toughness values emphasizes the "objectiveness" of K_{Ic} obtained by the compliance method (section 5.3.2). According to LEFM, Eq. 3 is only valid if no energy is dissipated outside the created surface of the crack which is assumed to be a straight line. However, in the determination of G_F, energy dissipated outside the crack surface is included, and the crack in concrete is not a straight line. Consequently, K_{Ic}^{LEFM} converted from G_F is theoretically larger than K_{Ic} evaluated by the compliance or alternative method, irrespective of the specimen size. According to size effects, the difference between K_{Ic}^{LEFM} and K_{Ic} decreases with increasing specimen size.

4.2 Evaluation of the tensile softening diagram

4.2.1 Numerical method

In order to obtain the tensile softening diagram from the fracture mechanics tests, the experimental F_s - COD mean curves of the WS Tests were numerically analysed using the program SOFTFIT (Roelfstra and Wittmann 1986). This program, which is based on the Fictitious Crack Model (Hillerborg et al. 1976), is used to determine a bilinear softening diagram from experimentally obtained force vs. COD curves by means of a data fit. The fracture test is simulated numerically using an initial (chosen) softening diagram, and, then, the calculated F_s vs. COD-curve is compared to the experimental curve. The parameters of the softening diagrams are corrected, and the analysis is repeated until good agreement between calculated and experimental curve is obtained. The softening diagram leading to this good agreement is considered to be the tensile softening diagram of the tested concrete.

The parameters of the bilinear softening diagram are the tensile strength f_t, the stress s_1 at the break point, the fictitious crack width w_1 at the break point and the maximum fictitious crack width w_2 according to Figure 10. A specific fracture energy value G_F, c is calculated from the area under the softening diagram.

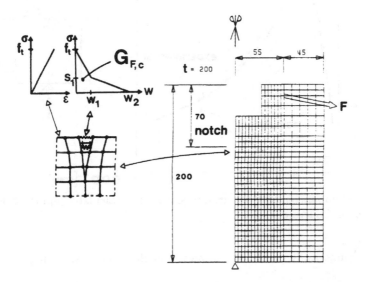

Figure 10: Schematical presentation of the bilinear softening diagram and finite element mesh used for the analysis.

Several parameters of the softening diagram were fixed in order to get unique solutions (Wittmann *et al.* 1988):

1. The tensile strength was fixed. The values of f_t were chosen taking into consideration the tensile splitting strength $f_{t,sp}$ of each concrete (Table 2).

2. The ratio of $\frac{s_1}{f_t}$ was fixed to a value of $\frac{1}{3}$. Concrete B showed a steep drop of the descending branch in the F_s-COD curve, and, a ratio of $\frac{s_1}{f_t} = \frac{1}{4}$ gave better agreement between experimental and calculated F_s-COD curves.

3. The fictitious crack width values w_1 and w_2 were determined by a data fit such that the square of the difference between experimental and calculated curves became minimal.

4.2.2 Results and discussion

Figure 11 shows F_s-COD curves for Concrete A after the data fit, chosen as an example to demonstrate the agreement of experimental and numerically predicted curves. The evaluated parameters of the bilinear softening diagrams are listed in Table 5 and presented graphically in Figure 12. The results of the numerical analysis can be discussed in the following way:

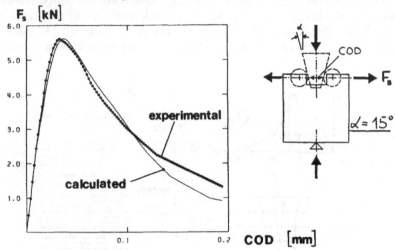

Figure 11: Comparison between experimental and calculated splitting force vs. COD curve for Concrete A.

Table 5: Numerically evaluated parameters of the bilinear softening diagrams.

WS SPECIMEN	CONCRETE	f_t [MPa]	s_1 [MPa]	w_1 [mm]	w_2 [mm]	$G_{F,c}$ [N/m]	$\frac{G_{F,c}}{G_F}$ [-]
CYLINDRICAL	A	3.6	1.2	0.0076	0.091	68.3	0.99
	B	4.4	1.1	0.013	0.064	63.8	0.93
	C1	3.6	1.2	0.011	0.064	58.2	0.97
	C2	3.0	1.0	0.0056	0.081	48.9	1.00
CUBIC	A	3.6	1.2	0.011	0.112	87.0	0.99
	B	4.4	1.1	0.016	0.067	72.1	0.95
	C1	3.6	1.2	0.014	0.073	69.0	0.95
	C2	3.0	1.0	0.0082	0.076	50.3	0.92

- Under the assumption of a bilinear softening diagram, force-COD curves can be calculated by means of finite element analysis leading to good agreement with the experimentally determined curves.

106

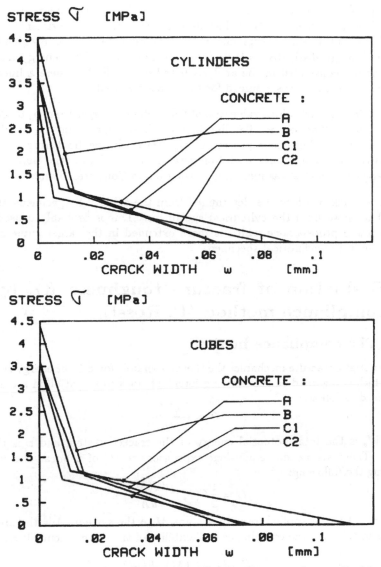

Figure 12: Numerically evaluated bilinear softening diagrams from cylindrical and cubical WS specimens.

- The $G_{F,c}$-values obtained from the calculated softening diagrams are up to 8 % smaller than the G_F-values determined from the experimental F_s-COD curves. This result is explained by the experimental G_F-value which includes also energy dissipated outside the fracture surface. This energy dissipation was not considered in the analysis based on the FCM, where a linear stress vs. strain curve was assumed for the intact material.

- Maximum crack width value, w_2, of Concrete A is larger than w_2 of Concretes B and C. The second, flat line in the bilinear curve may be attributed to the pull-out behaviour process of aggregates during fracturing. The more tortuous fracture surface of Concrete A indicated that the pull-out of the larger aggregates was more important than in Concretes B and C.

- The crack width values determined from the cubic WS specimens are larger than those from the cylindrical specimens. This is probably caused by the fracture process zone which is more extended in the larger cubic specimen than in the cylindrical specimen.

5 Evaluation of fracture toughness K_{Ic} by the compliance method (P. Rossi)

5.1 The compliance method

In linear elastic fracture mechanics the thermodynamic force G, or energy restitution ratio which is regarded as the "driving force" of crack propagation in a material, is introduced as follows:

$$G = -\frac{\delta W_p}{\delta a} \tag{4}$$

where W_p is the total potential energy of the cracked system and a is the crack length. There are various equivalent expressions for G of a linear elastic system including the following:

$$G = \frac{1}{2}F(M)^2 \frac{dC(M)}{dA} \tag{5}$$

where $F(M)$ is the applied force to a point M of the system, A is the area of the crack, and $C(M)$ is the compliance or flexibility of the system defined by:

$$U(M) = C(M) \cdot F(M) \tag{6}$$

with $U(M)$ being the displacement at point M. In LEFM, the stress intensity factor K is introduced as a factor that governs the stress singularity at the crack tip. Although G and K are taken from different approaches, both the energetic and analytical approach can yield criteria of crack propagation in a linear elastic medium. For crack propagation Mode I follows:

- according to Griffith criterion (Griffith 1921)

$$G_I = G_{Ic} \tag{7}$$

- and according to Barenblatt's criterion (Barenblatt 1962)

$$K_I = K_{Ic} \tag{8}$$

where G_{Ic} and K_{Ic} are the critical energy restitution ratio and the critical stress intensity factor, also called fracture toughness, respectively. K_{Ic} is related to G_{Ic} as derived by Irwin (1957):

$$G_{Ic} = \frac{K_{Ic}^2}{E'} \tag{9}$$

where,

$$E' = \begin{cases} E & : \text{for plane stress} \\ \frac{E}{(1-\nu^2)} & : \text{for plane strain} \end{cases} \tag{10}$$

Equation 9 substituted into equation 5 yields:

$$K_I = (\frac{E}{2}F(M)^2\frac{dC(M)}{dA})^{\frac{1}{2}}: \text{for plane stress} \tag{11}$$

$$K_I = (\frac{E}{2(1-\nu^2)}F(M)^2\frac{dC(M)}{dA})^{\frac{1}{2}}: \text{for plane strain} \tag{12}$$

Equations 11 and 12 show that it is possible to calculate K_{Ic} if the crack propagates and the critical length a_c of this crack, the critical force $F_c(M)$, the specimen compliance $C(a)$ and the derivative of $C(a)$ at point a_c are known. This method of determination of K_{Ic} is called the compliance method. In order to apply this method, the analytical expressions for C as function of a must be evaluated first for each specimen considered (see the following section).

The experimental determination of the critical crack length for concrete a_c is difficult because the crack visible at the surface is not representative of the real intra-volume crack (Rossi et al. 1986, Bascoul et al. 1987) for a number of reasons such as:

- drying shrinkage that induces "skin" microcracking,

- the heterogeneity of the concrete between the volume and the surface,

- the change from plane strain behaviour in the volume to plane stress behaviour at the surface.

Consequently, crack length measurements at the surface using optical methods are not reliable. To solve this problem, the notion of effective crack length is introduced. The effective crack, a_{eff}, is an idealized crack which is determined by assuming at a given point on the experimental force vs. COD curve, a_{eff} leads to the same specimen compliance as the real crack consisting of an irregular front preceded by the fracture process zone. The effective crack length is calculated using the following equation:

$$C_{exp} = C(a)_{theo} \tag{13}$$

where C_{exp} is the experimental compliance calculated at each loading/unloading cycle, and $C(a)_{theo}$ is the analytical expression obtained by the finite element analysis. Figure 13 illustrates this method of crack length determination.

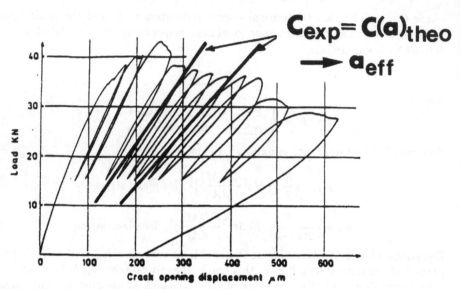

Figure 13: Compliance method determination of an effective crack length.

The specimen compliance is determined by taking the tangential slope of the reload curve. Significant residual stresses occur in concrete at the initial state due to phenomena of shrinkage. As a consequence, during propagation of both macrocrack and microcracks (fracture process zone), local release of a part of these initial stresses, and, hence, the development of residual deformation is observed:

- During unloading fretting forces between microcrack faces prevent the complete closing of these microcracks and the macrocrack. This phenomenon leads to a significant residual macroscopic deformation which can be observed in all fracture tests on concrete.

- During reloading, additional microcracks appear in the microcracked zone in the vicinity of the macrocrack. This implies that forces occurring at the beginning of the unloading do not coincide with those occurring at the end of the reloading.

5.2 Finite element analysis

The same procedure was followed for both TDCB and WS specimens. Four independent calculations were performed for four different fixed crack lengths. The material (concrete) is assumed to behave in a linear elastic way. One compliance value per calculation (or crack length) is obtained, and the equation for $C(a)$ is found by linear regression using the four values of C corresponding to the four lengths. Table 6 gives the crack lengths used for the calculations of the three specimens. Figure 14

Table 6: Crack lengths used for the FE-analysis.

Crack lengths in [m]				
TDCB	0.55	0.75	1.25	1.55
WS CYLINDER	0.047	0.052	0.070	0.090
WS CUBE	0.059	0.070	0.110	0.150

shows the meshes with their boundary conditions as used. For reasons of symmetry, only the mesh of a half-specimen is considered in each case.

For both WS specimens, the force acting on the bearings is split up into two components: a vertical force F_v and a horizontal splitting force F_s according to Figure 14. Both force components were considered in the numerical analysis using the following equation:

$$F_s = \frac{F_v}{tan\alpha} \tag{14}$$

where α is the wedge angle as shown in Figure 3. From the numerical analysis, the following two relations are obtained using the theorem of reciprocity of Betti:

$$U(a) = C_H(a) \cdot F_s + \nu(a) \cdot F_v \tag{15}$$

and

$$V(a) = C_V(a) \cdot F_v + \nu(a) \cdot F_s \tag{16}$$

where $U(a)$ and $V(a)$ are respectively the horizontal displacement (COD) and the vertical displacement at the point where F_s and F_v are applied. Equations 15 and 16 are rearranged using equation 14:

$$U(a) = C_h(a) \cdot F_s \tag{17}$$

$$V(a) = C_v(a) \cdot F_v \tag{18}$$

where:

$$C_h(a) = (C_H(a) + \nu(a) \cdot tan\alpha) \tag{19}$$

$$C_v(a) = (C_V(a) + \frac{\nu(a)}{tan\alpha}) \tag{20}$$

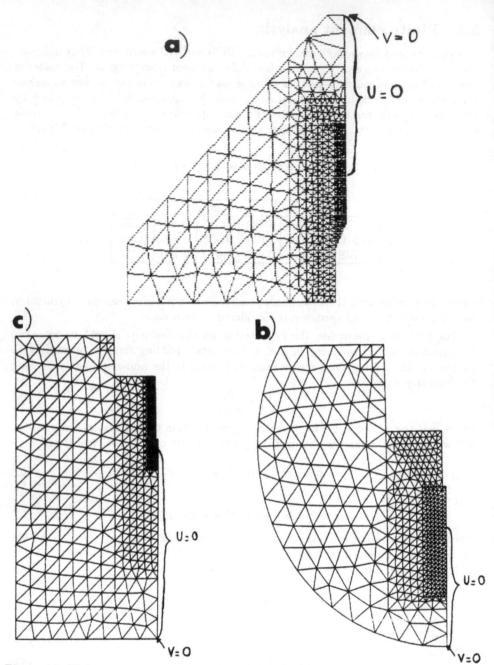

Figure 14: Finite element mesh of a) the TDCB, b) the cylindrical WS, and c) the cubical WS specimen.

where $U(a)$, F_s and F_v are experimentally determined. Relations 13 and 17 are considered to evaluate the effective crack length. Since the compliances of all specimens are calculated on the assumption of linear elasticity, they are inversely proportional to the Young's modulus E of the material. In all calculations, a value of $E = 3.5$ 10^{10} Pa was used. Consequently, the resulting functions are multiplied by $\frac{3.5 \cdot 10^{10}}{E}$ so that these functions are valid for any value of E. Finally, the following analytical expressions were obtained for the specimen compliances:

- TDCB specimen:

$$C(a) = \frac{1}{E}(3410.75a^3 - 7470.05a^2 + 5700.1a - 1281) \ [m/N] \qquad (21)$$

for 0.55 m < a < 1.55 m

- cubic WS specimen:

$$C(a) = \frac{1}{E}(268.1a^3 - 58.7a^2 + 4.4a - 0.1) \cdot 10^4 \ [m/N] \qquad (22)$$

for 0.059 m < a < 0.15 m

- cylindrical WS specimen:

$$C(a) = \frac{1}{E}(33.03a^3 - 5.19a^2 + 0.28a - 0.005) \cdot 10^6 \ [m/N] \qquad (23)$$

for 0.047 m < a < 0.090 m

where a is measured from the point where F_s and F_v are applied (Fig. 14). Equations 11, 21, 22 and 23 yield the following expressions for the stress intensity factor $K_I(a)$ according to the calculations in plane stresses:

- TDCB specimen:

$$K_I(a) = \frac{1}{\sqrt{2B}}(10232.25a^2 - 14940.1a + 5700.1)^{1/2} \cdot F \ [MPa\sqrt{m}] \qquad (24)$$

for 0.55 m < a < 1.55 m

- cubic WS specimen:

$$K_I(a) = \frac{1}{\sqrt{2B}}(804.3a^2 - 117.4a + 4.4)^{1/2} \cdot 10^2 \cdot F_s \ [MPa\sqrt{m}] \qquad (25)$$

for 0.059 m < a < 0.15 m

- cylindrical WS specimen:

$$K_I(a) = \frac{1}{\sqrt{2B}}(99.09a^2 - 10.38a + 0.28)^{1/2} \cdot 10^3 \cdot F_s \ [MPa\sqrt{m}] \qquad (26)$$

for 0.047 m < a < 0.090 m

where B is the specimen thickness.

5.3 Values of K_{Ic} obtained in the two types of test

5.3.1 Evaluation of effective Young's modulus and fracture toughness

Equations 21 - 23 contain the Young's modulus E. For the calculation of the effective crack length, the "true" Young's modulus of each concrete must be known. However, many studies have shown that the Young's modulus determined on standard specimens does not necessarily correspond to E obtained from a structure. For this reason, the notion of effective Young's modulus E_{eff} is introduced as the value determined from each specimen tested:

$$C(a_0)_{theo} = C_{exp} \qquad (27)$$

where $C(a_0)_{theo}$ is the calculated theoretical compliance for the initial "crack" length a_0 which is the notch length. C_{exp} is the experimental compliance determined during the first loading in the linear elastic domain before the formation of the microcracked zone at the notch tip. Using equation 27, the value of E_{eff} is determined from each specimen and concrete, and E in equations 21 - 23 is then replaced by E_{eff} in order to obtain the effective crack length a_{eff}. The fracture toughness K_{Ic} of each specimen and concrete is then calculated using equations 24 to 26, a_{eff} and the corresponding maximum force value of each unload/reload cycle.

5.3.2 TDCB tests

The curves of fracture toughness as function of the effective crack length are plotted on Figure 15, and, Table 7 gives the mean K_{Ic}-values obtained for the three concretes tested. These results suggest the following remarks:

Table 7: K_{Ic}-values obtained from TDCB tests.

CONCRETE	K_{Ic} [MPa$\sqrt{m}$]
A	2.21
B	2.11
C2	1.31

- The K_{Ic}-values as function of a_{eff} fluctuate between a minimum and a maximum value. This may be caused by the heterogeneity of concrete where a crack encounters hard spots (aggregates) and weak spots (flaws) in a random manner. The fluctuation increases as the size of the largest aggregates increases which is consistent with the foregoing explanation.

114

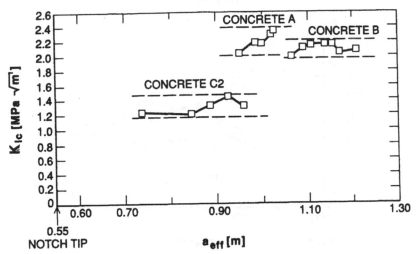

Figure 15: Fracture toughness as function of effective crack length: TDCB Tests.

- The mean value of K_{Ic} increases with growing aggregate size, which is explained assuming that K_{Ic} depends primarily on the energy consumed in the microcracked zone (fracture process zone) leading the macrocrack. This energy is related to the size of the largest aggregates.

- The mean value of K_{Ic} for Concrete B with a maximum aggregate size of 12 mm is very close to the values found by the LCPC (Rossi 1986, Rossi *et al.* 1986) on a similar concrete with the same grading, in tests carried out on large Double Cantilever Beam specimens having a geometry different from the TDCB. This finding indicates that objective K_{Ic}-values independent of specimen size and shape were determined using TDCB specimens.

5.3.3 Wedge splitting tests

Figure 16 shows representative curves for the fracture toughness as function of the effective crack length for the three concretes and the two types of WS specimens. All curves have the same general shape, i.e., K_{Ic} first increases with increasing crack length, and decreases in the second part of the curve. There is no "plateau" where K_{Ic} as a function of a_{eff} is constant, except for the curve relative to Concrete C2 tested on cubic specimens. The shape of these curves can be explained as follows:

- The increase of K_{Ic} in the first part of the curve may be attributed to the increasing size of the microcracked zone at the crack tip, which is the stage of formation of the fracture process zone, a transitional crack propagation regime.

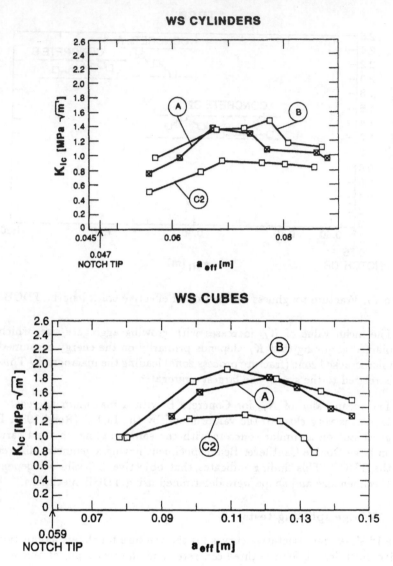

Figure 16: Fracture toughness as function of effective crack length: Wedge Splitting Tests.

- The sudden decrease of K_{Ic} in the second part of the curve is explained by the fact that there is an interaction between the conditions at the geometrical boundaries, i.e., compressive stresses at the free edges of the specimen, and the fracture process zone. Since the fracture process zone cannot develop fully under these geometrical conditions, part of the energy that would have dissipated in the fracture process zone is dissipated in crack propagation resulting in the decrease of K_{Ic}. This stage also involves a transitional crack propagation regime.

- The fact that no plateau appears on the curves relative to Concretes A and B indicates that no steady crack propagation regime was reached. Consequently, with reference to the basic assumptions of linear elastic fracture mechanics, the tests on the WS specimens do not yield "objective" K_{Ic}-values for these two concretes. However, a plateau is reached for Concrete C2 with the cubic specimen indicating that the "objective" K_{Ic} has been determined for this concrete.

- The effective crack length values are smaller than the values practically possible, i.e., the specimen height. An explanation for this observation is based on the inherent assumptions of the compliance method. A perfectly linear elastic system was assumed in the finite element calibration resulting in zero residual CMOD upon unloading, but the assumption was increasingly violated as the test progressed. As the effective crack length increased, increased residual CMOD (Fig. 7) caused by the inability of the crack perfectly to close was observed. This effect becomes predominant as the crack reaches a certain length. The same phenomena was observed in Saouma et al. (1989).

Consequently, only the maximum fracture toughness values were considered. The single and mean values with the relative standard deviation, s, are summarized in Table 8.

Table 8: K_{Ic}-values obtained from TDCB tests.

WS Specimen	Concrete	K_{Ic} [MPa$\sqrt{m}$]					$K_{Ic,mean}$ [MPa$\sqrt{m}$]	s [%]
Cylindrical	A	1.38	1.51	1.14	1.72	-	1.44	16.9
	B	1.43	1.67	1.48	1.48	-	1.52	7.0
	C2	0.92	1.10	0.94	-	-	0.99	10.0
Cubic	A	1.95	1.91	1.61	1.82	1.83	1.82	7.2
	B	1.94	1.97	2.11	1.68	-	1.97	9.5
	C2	1.31	1.35	1.50	-	-	1.39	7.2

Table 8 suggests the following remarks:

- For all concretes tested, K_{Ic} determined on the cubic specimens is higher than K_{Ic} obtained from the cylindrical specimens. This result is explained by the longer ligament length of the cubic specimen allowing a crack propagation régime which is closer to the steady crack propagation régime.

- Both types of specimen yield K_{Ic}-values for Concrete A which are lower than those found for Concrete B, which is inconsistent with findings for large TDCB specimens (Table 7). The contradiction is explained as follows. The dimensions of the fracture process zone may depend on the size of the largest aggregate, (that is, the larger the diameter, the larger the fracture process zone), and, the interaction between compressive stresses and the free edge of the specimen and the microcracked zone may be more important for Concrete A than for Concrete B. On the other hand, the larger K_{Ic}-value for Concrete B is consistent with the strength properties of the tested concretes (Table 2) where both f_{cc} and f_t of Concrete B are larger than the corresponding values of Concrete A.

- The K_{Ic}-values obtained for Concrete C2 are small, because of the poor quality of Concrete C2 (Table 2) due to bad compacting.

5.4 Determination of the equation linking the values of K_{Ic} obtained in the TDCB and WS tests

Since for common concrete the fracture toughness may be directly related to the size of the largest aggregates, a relationship between the ratio of $\frac{K_{Ic}(WS)}{K_{Ic}(TDCB)}$ and the size of the largest aggregates d_a was established. This ratio is reported in Table 9 and plotted in Figure 17 for all concretes investigated and for both WS specimen shapes:

Table 9: Comparison of K_{Ic}-values as obtained from the TDCB test and both types of WS specimen.

$\frac{K_{Ic}(WS)}{K_{Ic}(TDCB)}$			
Concrete	A	B	C2
WS Cylinder	0.65	0.72	0.76
WS Cube	0.82	0.93	1.00

- The smaller the ratio of maximum aggregate size to specimen size, the more "objective" is K_{Ic} of plain concrete, which is consistent with the literature on size effect.

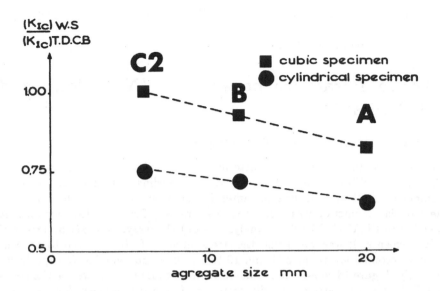

Figure 17: Ratio of $\frac{K_{Ic}(WS)}{K_{Ic}(TDCB)}$ as function of the maximum aggregate size.

- The specimen thickness may have an influence on K_{Ic}, because K_{Ic} obtained on the cubical WS specimen is not very different to K_{Ic} obtained on the TDCB specimen even though the plan dimensions are very different. This may be explained by the energy dissipation in the fracture process zone which may be a volume energy dissipation, and, consequently, the thicker the specimen, the smaller the plan dimensions of the fracture process zone, which improves the confinement conditions of the fracture process zone and the applicability of LEFM. "Objective" K_{Ic}-values of plain ~oncretes might possibly be determined on slightly larger cubical WS specimens. Using the compliance method, "objective" K_{Ic}-values were determined on WS specimens with a height of 300 mm (Saouma *et al.* 1989).

- For Concrete C2, K_{Ic}-values as obtained with TDCB and WS cubes are the same confirming the results described in section 5.3.2.

- For a maximum aggregate size ranging from 6 to 20 mm, linear relations between the ratio of $\frac{K_{Ic}(WS)}{K_{Ic}(TDCB)}$ and the aggregate size d_a can be determined. These relations depend on the specimen shape. The following equations are proposed:

for the cubical WS specimen: $\dfrac{K_{Ic}(WS)}{K_{Ic}(TDCB)} = -0.075\dfrac{d_a}{d_a^*} + 1.075$ \qquad (28)

119

for the cylindrical WS specimen: $\dfrac{K_{Ic}(WS)}{K_{Ic}(TDCB)} = -0.042\dfrac{d_a}{d_a^*} + 0.802$ (29)

where $d_a^* = 6$ mm and 6 mm $< $ d $ < 20$ mm.

6 Mode I fracture parameters (Y.S. Jenq and S.P. Shah)

6.1 Evaluation of fracture toughness

Finite element methods were used to determine the compliance and the stress intensity factor of the WS specimens. Contrary to the preceding section where the classical compliance method with the stress intensity factor obtained from the derivation of the compliance function was used, the stress intensity factors in this section were obtained from FRANC (FRacture Analysis Code) (Wawrzynek 1987) by means of singular elements. It was found that the stress intensity factor induced by the vertical force accounts for about 3 % and 10 % for cube and cylinder WS specimens respectively. Figure 18 shows the evaluated fracture toughness values for Concretes A and B. The critical stress intensity factor, or fracture toughness K_{Ic}, is essen-

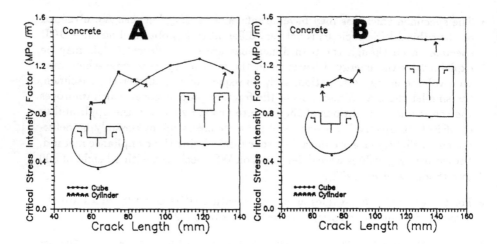

Figure 18: Critical stress intensity factor after crack propagation for   Concrete A, and,   Concrete B.

tially constant after peak load. However, K_{Ic} obtained from the cube specimen was always larger by 10 to 20 % than K_{Ic} of the cylindrical specimens. Comparison with K_{Ic}-values determined by Rossi (Section 5) depicts some discrepancies, i.e.,

K_{Ic}-values evaluated using FRANC are about 30 % smaller for both Concretes A and B. These differences are explained as follows:

- Only four crack lengths were used by Rossi to curve fit the compliance function, and, significant errors may be introduced in the derivation of such a compliance function to obtain the stress intensity function. Comparison of the two stress intensity functions (Fig. 19) shows that the stress intensity function determined by Rossi has a minimum value at crack length of 82 mm,

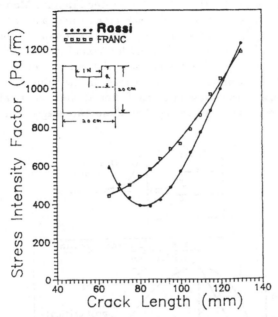

Figure 19: Stress intensity factor obtained using compliance method and singular element method - Cube WS specimen.

which was not observed from the singular element approach. This observation is believed to be due to lack of data points used in the calculation of the compliance function. Similar discrepancies were also observed for cylindrical and TDCB specimens.

- The unloading elastic compliance determined in the present approach is different from that used by Rossi in which the tangential slope of the reloading curve is used. In the present approach, the unloading point to the final reloading point is used. Due to different approaches used in determining the unloading compliance, the effective crack length determined by Rossi (sections 5.2 and 5.3) is longer than that determined using the unloading compliance determined from the present approach. In addition, the stress intensity factor obtained using FRANC for the same applied load is lower than that determined from the compliance method as indicated in Figure 19 for crack lengths ranging from 95 mm to 140 mm. As a result, the fracture toughness values determined by Rossi were found to be higher (by about 40 % - 60 %) than those determined using the method reported in this section.

6.2 Fracture parameters according to the Two Parameter Fracture Model

In the Effective Griffith Model, or Two Parameter Fracture Model, two stages are assumed for crack growth: precritical crack growth and postcritical crack growth (Fig. 20). Assuming that a crack will propagate at a constant value of stress intensity factor (K_{Ic}^s, in which s denotes the consideration of stable crack growth) when it reaches the critical point. Before the critical point is reached, the precritical stable crack growth will be accompanied by an increase in the value of the stress intensity factor. The increase of the stress intensity factor at the precritical region is known as the R-curve behaviour. Assuming that these two stages can be separated by

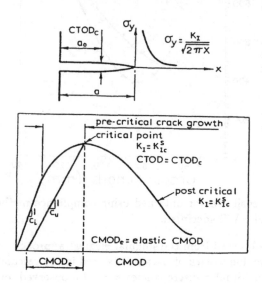

Figure 20: Possible stages of crack propagation in plain concrete.

a well-defined critical point. This critical point is defined by two fracture parameters: critical stress intensity factor, K_{Ic}^s, and critical crack tip opening displacement, $CTOD_c$. Based on the definition of the critical point, the stress intensity factor calculated using the effective crack length (which is the sum of the notch length, a_0, and the critical crack extension), and the applied critical load should be equal to K_{Ic}^s at this point. Thus, the critical effective crack length and the applied critical load can be calculated by satisfying the following conditions: the stress intensity factor should be equal to K_{Ic}^s at the effective crack tip, a, and the crack tip opening displacement measured at the notch tip, a_0, is equal to $CTOD_c$ (Fig. 20). Depending on the specimen geometry, the critical load may not be the peak load. It can be shown that the peak load is the same as the critical load for three-point bend configuration (Jenq and Shah 1985).

The two fracture parameters associated with the effective Griffith model can be determined on fracture specimens by assuming that the difference between the initial compliance, C_i, and the unloading compliance at peak load, C_u, is mainly due to the formation of critical crack growth. Therefore, knowing the ratio of unloading compliance and initial compliance, the effective crack length can be determined using iteration scheme detailed in (Jenq and Shah 1985, Karihaloo and Nallathambi 1989). Knowing the notch length and compliance ratio, R_c, the effective crack length can be determined. From the effective crack length and the measured peak load, P_{max}, the critical stress intensity factor and critical crack tip opening displacement can then be determined using available LEFM formulae or by finite element calibration for specimens with different configurations.

The two fracture parameters K_{Ic}^s and $CTOD_c$ according to the Two Parameter Fracture Model (Jenq and Shah 1985) are given in Table 10 for cubical and cylindrical specimens. The reported K_{Ic}^s and $CTOD_c$ values were determined from the measured peak load and compliance ratios. From the results, it can be noticed

Table 10: Fracture parameters according to the Two Parameter Model for Concrete A and B.

WS SPECIMEN	CONCRETE	K_{Ic}^s [MPa$\sqrt{m}$]					$K_{Ic,mean}^s$ [MPa$\sqrt{m}$]	s [%]
CYLINDRICAL	A	0.94	0.92	0.76	1.06	-	0.92	13.4
	B	1.00	1.24	1.10	1.11	-	1.11	8.9
CUBIC	A	1.08	1.27	0.93	1.12	1.12	1.10	11.0
	B	1.36	1.50	26	1.36	-	1.37	7.2
WS SPECIMEN	CONCRETE	$CTOD_c$ [mm]					$CTOD_{c,mean}$ [mm]	s [%]
CYLINDRICAL	A	0.0056	0.0056	0.0054	0.0081	-	0.0062	20.7
	B	0.0082	0.0108	0.0071	0.0062	-	0.0081	24.6
CUBIC	A	0.0118	0.0134	0.0072	0.0136	0.0142	0.0120	23.7
	B	0.0149	0.0129	0.0113	0.0133	-	0.0131	11.3

that larger K_{Ic}^s of cubic and cylindrical specimens are similar. The corresponding $CTOD_c$-values are not similar. Moreover, the differences in $CTOD_c$-values are comparable to those observed in G_F-values and reported in Table 3. There are several possible reasons for such geometry dependence as summarized below (Jenq and Shah

1989):

1. The values of $CTOD_c$ are rather small and unless carefully instrumented to avoid parasitic deformations, large errors may result. One way to check th accuracy of the measuring system is to calculate the modulus of elasticity value from the linear portion of the load vs. $CMOD$ curve and using the relevant LEFM equation. When this was done from the measurements of the data for the cubic and cylindrical curve wedge loaded specimens, the calculated E values were much lower than expected for concrete, and they were different for cubic and cylindrical specimens. This might perhaps explain the observed differences in $CTOD_c$ values. It is desirable to use as small a gage length as possible (for example, using clip gages, as recommended by ASTM for metals).

2. Effects of self-weight may also have significant impact on the determination of fracture parameters, especially for very large specimens. As the specimen size increases, the effect of self-weight of specimens becomes more important in the determination of initial compliance, unloading compliance, and maximum load. As a result, effects of self-weight should be included in the determination of fracture parameters of large specimens.

3. Very often the same loading device is used for different sizes of specimens throughout the same experimental program. Therefore, it is possible that the degree of alignment may also depend on the size of the specimens.

4. The accuracy of theoretical (or numerical) solutions of stress intensity factor, the associated displacments, and other physical variables is also very important in the determination of fracture parameters. For example, the determination of $CTOD_c$ is very sensitive to the accuracy of the determination of initial and unloading compliance as well as the accuracy of the available numerical (or theoretical) solutions of the crack tip opening displacement.

5. The ratio of specimen stiffness and machine stiffness is dependent on the specimen size. Depending on the testing configuration, the stability condition may be different for different specimen sizes.

Unless all the above non-material related factors can be excluded, it is very difficult to conclude that the observed size effect reported here is a true reflection of the material fracture property. More research effort is needed to resolve this problem. However, before this size effect problem is fully resolved, it is recommended that fracture parameters in the Two Parameter Fracture Model (i.e., K_{Ic}^s and $CTOD_c$) are determined using medium size specimens where the self-weight of the specimen is less than 30 % of the applied maximum load and the uncracked ligament length is at least three times of the maximum aggregate size.

There are two advantages associated with the use of medium size specimens. First, the specimens can be handled easily in the laboratory. Secondly, the fracture resistance of a concrete structure predicted using the two fracture parameters obtained from smaller specimens was found to be more conservative than those derived from large specimens. Therefore, from safety consideration, the fracture parameters obtained from smaller specimens shall be used to design a load bearing concrete structure even if there really exists an intrinsic effect of spcimen size on the fracture parameters.

7   Conclusions and recommendations

1. The Wedge Splitting Test is a promising test method for the determination of fracture mechanics properties. Simple specimen geometries like cubes and cylinders (drilled from existing structures) with a large fracture area compared to the specimen weight can be tested. The Wedge Splitting Test can be performed in a "normally equipped" civil and materials engineering laboratory. This testing method meets the requirements for an alternative method to the standardized three point bend test.

2. Tests on Tapered Double Cantilever Beams with large dimensions are necessary for the investigation of the influence of specimen size and maximum aggregate size on fracture properties of concrete.

3. Fracture properties obtained using different fracture specimens may be affected by the inherent assumptions in the evaluation of fracture properties, and by the test method itself. Fracture properties evaluated from test data must be interpreted in close conjunction with the fracture model considered.

4. Fracture properties according to the Fictitious Crack Model: The specific fracture energy G_F depends on both maximum aggregate size of the investigated concrete and the specimen dimension. G_F increases with both increasing maximum aggregate size and ligament length of the WS specimens. A bilinear tensile softening diagram can be obtained by a data fit procedure.

5. Fracture properties according to the Compliance Method: The ratio of largest aggregate size to the dimensions of the specimen significantly influences the determination of fracture toughness of concrete. In order to obtain "objective" K_{Ic}-values of plain concrete on the basis of tests on small specimens, the following equations are proposed:

for the cubical WS specimen: $\dfrac{K_{Ic}(WS)}{K_{Ic}(TDCB)} = -0.075\dfrac{d_a}{d_a^*} + 1.075$ (30)

for the cylindrical WS specimen: $\dfrac{K_{Ic}(WS)}{K_{Ic}(TDCB)} = -0.042\dfrac{d_a}{d_a^*} + 0.802$ (31)

where $d_a^* = 6$ mm and 6 mm $< d < 20$ mm.

6. Fracture properties according to the Two Parameter Model: Comparing the results of the cubic and the cylindrical specimens, it was observed that K_{Ic}^s-values were similar but $CTOD_c$ were not. The differences in $CTOD_c$-values were comparable to that observed in G_F-values. These differences can be due to possible errors in measuring rather small values of crack mouth opening displacement. Careful attention should be given to crack mouth opening displacement.

For future work, Subcommittee B recommends subjecting the Wedge Splitting Test to a more thorough investigation by analysing all available test data for concrete, mass concrete, other cementitious materials and rock. Additionally, other "compact" specimen testing methods should be evaluated and compared for use as a standardized testing method. Finally, standardized fracture tests used for rock or ceramics should be considered in order to obtain a broad consensus for a new standardized testing method for concrete.

8 References

Barenblatt, D.I. 1962. The mathematical theory of equilibrium cracks in brittle fracture. in "Advances in Applied Mechanics", Academic Press, pp. 55-129.

Bascoul, A., F. Kharchi,I. Maso. 1987. Concerning the measurement of fracture energy of a microconcrete according to the crack growth in a three point bending test on notched beams. Proceedings of the Intl. Conf. on Fracture of Concrete and Rock, Houston, pp. 631-643.

Bažant, Z.P. 1984. Size effect in blunt fracture: Concrete, rock, metal. J. of Engineering Mechanics, ASCE, 110(4), pp. 518-535.

Bažant, Z.P. and B.H. Oh. 1983. Crack band theory for fracture of concrete. Materials and Structures, Vol. 16, pp. 155-177.

Brühwiler, E. 1988. Fracture mechanics of dam concrete subjected to quasi-static and seismic loading conditions. Laboratory for Building Materials (LMC), Swiss Federal Institute of Technology, Lausanne, Thesis No. 739. (in German)

Brühwiler, E., J.J. Broz and V.E. Saouma. 1989. Fracture properties of dam concrete. Part III: Model assessment. Journal of Engineering Materials, ASCE, (submitted for publication).

Brühwiler, E. and F.H. Wittmann. 1990. The wedge splitting test, a method of performing stable fracture mechanics tests. Engineering Fracture Mechanics, Vol. 35, No 1/2/3, pp. 117-126.

Entov, V.M. and V.I. Yagust. 1975. Experimental investigation of laws of governing quasi-static development of macrocracks in concrete. I. 2V. RN.SSSR. Mekhniha, Tresdays Tela, Vol. 10, No 4, pp. 93-103.

Griffith, A.A. 1921. The phenomena of rupture and flow in solids. Phil. Trans. Roy. Soc. London, A221, pp. 163-197.

Hillemeier, B. 1976. Fracture mechanics investigation of crack propagation in cementitious materials. Doctoral Thesis, TU Karlsruhe. (in German)

Hillemeier, B. and H.K. Hilsdorf. 1977. Fracture mechanics studies on concrete compounds. Cement and Concrete Research, 7, pp. 523-536.

Hillerborg, A., M. Modeer and P.E. Petersson. 1976. Analysis of crack formation and crack growth in concrete by means of fracture mechanics and finite elements. Cement and Concrete Research, Vol. 6, pp. 773-782.

Hordijk, D., J. van Mier and H.W. Reinhardt. 1989. Material properties. in "Fracture Mechanics of Concrete Structures - From Theory to Application", edited by L. Elfgren, State of the Art Report, RILEM TC 90-FMA, Chapman & Hall, London, pp. 76-97.

Irwin, G.R. 1957. Analysis of stress and strains near the end of crack traversing plate. J. of Applied Mechanics, pp. 720-733.

Jenq, Y.S. and S.P. Shah. 1985. Two parameter fracture model for concrete. Journal of Eng. Mech. Div., ASCE, Vol. 111, No 10, pp 1227-1241.

Jenq, Y.S. and S.P. Shah. 1989. Geometrical effects on mode I fracture parameters. Proceedings, in "Fracture Toughness and Fracture Energy - Test Methods for Concrete and Rock", edited by H. Mihashi et al., Balkema, Rotterdam.

Karihaloo, B.L. and P. Nallathambi. 1989. Notched beam test: mode I fracture toughness. in "Fracture Mechanics of Concrete: Test Methods.", State of the Art Report, RILEM TC 89-FMT, Chapman & Hall, London.

Linsbauer, H.N. and E.K. Tschegg. 1986. Fracture energy determination of concrete with cube-shaped specimens. Zement und Beton, 31, 38-40. (in German)

Mindess, S. (1989). "Fracture process zone detection", Report of Subcommittee E, RILEM TC 89-FMT.

Planas, J. and M. Elices. 1989. Conceptual and experimental problems in the determination of the fracture energy of concrete. Proceedings, in "Fracture Toughness and Fracture Energy - Test Methods for Concrete and Rock", edited by H. Mihashi et al., Balkema, Rotterdam.

Regnault, P. and E. Brühwiler. 1989. Holographic interferometry for the determination of fracture process zone of concrete. Engineering Fracture Mechanics, Vol. 35, No 1/2/3, pp. 29-38.

RILEM Draft Recommendation (50-FMC). 1985. Determination of the fracture energy of mortar and concrete by means of three-point bend tests on notched beams. Materials and Structures, Vol. 18, pp. 287-290.

Roelfstra, P.E. and F.H. Wittmann. 1986. Numerical method to link strain softening with failure of concrete. in "Fracture Toughness and Fracture Energy of Concrete", edited by F.H. Wittmann, Elsevier, pp. 163-176.

Rossi, P. 1986. Cracking of concrete, from the material to the structure: Application of linear elastic fracture mechanics. Doctoral Thesis, Ecole National des Ponts et Chaussees. (in French)

Rossi, P., O. Coussy, C. Boulay, P. Acker, and Y. Malier. 1986. Comparison between plain concrete toughness and steel fiber reinforced concrete toughness.

Cement and Concrete Research, Vol. 16, pp. 303-313.

Saouma, V.E., J.J. Broz, E. Brühwiler and H.L. Boggs. 1990. Fracture properties of dam concrete. Part I: Laboratory Experiments. Journal of Engineering Materials, ASCE, (submitted for publication).

Wawrzynek, P.A. 1987. Interactive finite element analysis of fracture process: an integrated approach. Department of Structural Engineering, Cornell University.

Wittmann, F.H., K. Rokugo, E. Brühwiler, H. Mihashi and P. Simonin. 1988. Fracture energy and strain softening of concrete as determined by means of compact tension tests. Materials and Structures, Vol. 21, pp. 21-32.

3 MIXED-MODE CRACK PROPAGATION IN CONCRETE

A. CARPINTERI
Politecnico di Torino, Torino, Italy
S. SWARTZ
Kansas State University, Manhattan, USA

1 Introduction

The influence of in-plane shear on the process of cracking of concrete structures is of great practical importance. In the terminology of fracture mechanics, this is called mode II or mixed-mode loading.

In the past few years, extensive efforts have been directed to modeling crack propagation and fracture in concrete and rock subjected to mixed-mode loadings. In this report the term "mixed-mode" implies the presence of in-plane, normal and shearing tractions in the immediate region of the front of an existing crack and behind the front as well. Thus the crack and its "tip" are subjected to combined stresses. The deformation at the crack tip may have both mode I (opening) and mode II (sliding) components.

1.1 Background

The concepts of stress concentration and stress intensification are introduced in a brief exposition of the development of the ideas of Linear Elastic Fracture Mechanics (LEFM). The crack size and structure size transitions are then discussed thus connecting the two limit cases of ultimate tensile failure and LEFM.

The inherent material defects are often considered as the principal cause of brittle fracture. On the other hand, the stress concentration effects in the vicinity of flaws and defects have been well-known for a long time. Already in 1898, Kirsch provided the solution to the problem of an infinite plate in tension with a circular hole. The maximum stress at the hole edge is three times higher than the external one (Fig. 1). This means that, even if the size of the hole is negligible with respect to the size of the plate, the strength of the plate is reduced to one third. The problem solution is a compromise between removed material and hole curvature. As a limit-situation, even an infinitesimal hole produces a stress concentration factor equal to 3. In fact, the radius of curvature of such a hole is nearly zero in this case and produces conditions of particular severity.

Inglis (1913), see Broek (1986) extended the stress concentration investigation to the more general case of an elliptical hole (Fig. 2). The maximum stress at the hole edge is multiplied by the factor (1+2a/b) in this case. The strength of the flawed plate depends only on the ratio of the semiaxes of the contour ellipse. The stress

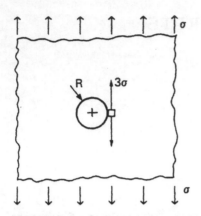

Fig. 1. Stress-concentration at the edge of a circular hole.

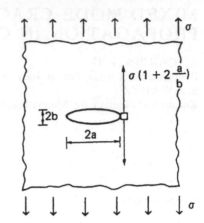

Fig. 2. Stress-concentration at the edge of an elliptical hole.

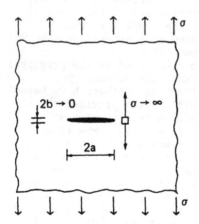

Fig. 3. Limit-case of a highly-prolated elliptical hole.

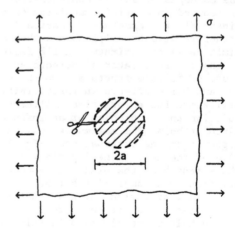

Fig. 4. Strain energy release in a uniformly stretched plate.

concentration factor increases by increasing the ellipse elongation. When $a/b \to \infty$, i.e., when the ellipse is very prolate, the stress concentration factor tends to infinity. Then, such a model does not appear to be useful to describe the critical condition for a notch of length 2a and vanishing width 2b (Fig. 3). A very small external load is sufficient, in fact, to exceed the ultimate tensile strength σ_u at the notch tips. On the other hand, from a practical point of view, the cracked bodies can sustain even considerable loads.

1.2 Griffith's Model

In 1920, Griffith (Broek, 1986) introduced energy-considerations (and not only stress-considerations) to explain the phenomenon of fracture. He proved that the elastic energy W_e released by a uniformly stretched plate of unit thickness, when the latter is cut by a crack of length 2a, is proportional to the energy contained in the circle of radius a prior to cutting (Fig. 4):

$$W_e = \pi \, a^2 \frac{\sigma^2}{E} \qquad (1)$$

where E is the elastic modulus of the material.

On the other hand, in order to produce a crack of length 2a, the following surface energy is necessary:

$$W_s = 2a \, G_{IC} \qquad (2)$$

where G_{IC} is the energy related to the unit crack area. Griffith assumed that, in order for an initial crack of length 2a to grow, the elastic energy released in a virtual crack extension must be larger than or equal to the energy required by the new portion of free surface:

$$\frac{dW_e}{da} \geq \frac{dW_s}{da} \qquad (3)$$

The instability condition reads as follows:

$$2 \, \pi \, a \, \frac{\sigma^2}{E} \geq 2 \, G_{IC} \qquad (4)$$

The preceding inequality holds for the applied stress σ as well as for the half-length of the crack a. The pairs of values σ and a under the curve in Fig. 5 represent stable cases, whereas pairs above represent unstable cases. The instability condition for the applied stress reads:

$$\sigma \geq \sqrt{\frac{E \, G_{IC}}{\pi \, a}} \qquad (5)$$

131

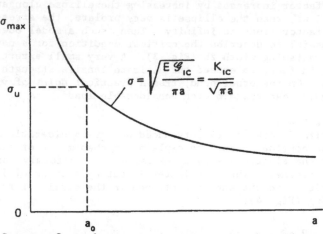

Fig. 5. Stress of crack propagation versus crack half-length (plate of infinite width).

1.3 Irwin's Model

After Griffith, fracture mechanics halted and was reconsidered only twenty years later. Particularly, on the occasion of very serious accidents which occurred to Liberty ships during the Second World War, the analysis of the fracture phenomena continued at a faster and faster rate. It was clear, in fact, that the material strength is not sufficient to guarantee the safety of such important structures. Other material properties, related to the formation of free surfaces (cracks), are required. Such properties are called "fracture toughness" and represent the energy dissipated in a unit crack growth.

After a decade of attempts to generalize the Griffith's model, a fundamental contribution was given by Irwin. He reconsidered the cracked plate subjected to uniform stress (Fig. 6) and firstly analysed it from the stress point of view. As was previously pointed out, the stress in the crack tip vicinity tends to infinity. Irwin defined the "rate" at which such a stress tends to infinity as the distance from the crack tip tends to zero. That is, he defined the stress-singularity power and found that such a power is always -1/2, independently of the structure and loading geometry. More precisely, Irwin proposed a mathematical crack model referring the near stress field to three fracture modes (Fig. 7). They are the opening, sliding and tearing modes and are often referred to respectively as mode I, II and III crack extension. The corresponding stress fields are:

$$\sigma_x = \frac{K_I}{(2\pi r)^{1/2}} \cos \frac{\theta}{2} \left[1 - \sin \frac{\theta}{2} \sin \frac{3}{2} \theta \right] \qquad (6\text{-}a)$$

$$\sigma_y = \frac{K_I}{(2\pi r)^{1/2}} \cos \frac{\theta}{2} \left[1 + \sin \frac{\theta}{2} \sin \frac{3}{2} \theta \right] \qquad (6\text{-}b)$$

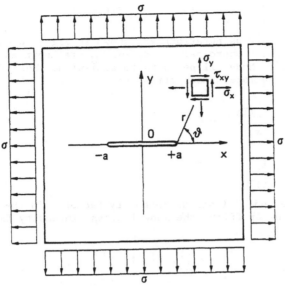

Fig. 6. Stress notation in the crack tip vicinity.

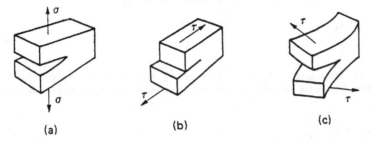

(a)   (b)   (c)

Fig. 7. Fracture modes: opening (a), sliding (b) and tearing (c).

$$\tau_{xy} = \frac{K_I}{(2\pi r)^{1/2}} \sin\frac{\theta}{2} \cos\frac{\theta}{2} \cos\frac{3}{2}\theta \qquad (6\text{-}c)$$

for mode I and

$$\sigma_x = -\frac{K_{II}}{(2\pi r)^{1/2}} \sin\frac{\theta}{2} \left(2 + \cos\frac{\theta}{2} \cos\frac{3}{2}\theta\right) \qquad (7\text{-}a)$$

$$\sigma_y = \frac{K_{II}}{(2\pi r)^{1/2}} \cos\frac{\theta}{2} \sin\frac{\theta}{2} \cos\frac{3}{2}\theta \qquad (7\text{-}b)$$

$$\tau_{xy} = \frac{K_{II}}{(2\pi r)^{1/2}} \cos \frac{\theta}{2} \left[1 - \sin \frac{\theta}{2} \sin \frac{3}{2}\theta \right] \tag{7-c}$$

for mode II. The stress notations are shown in Fig. 6 with r and θ being the local polar coordinates. K_I and K_{II} are known as the stress intensity factors. Mode III pertains to anti-plane shear and is associated with the local shear stresses:

$$\tau_{xz} = - \frac{K_{III}}{(2\pi r)^{1/2}} \sin \frac{\theta}{2} \tag{8-a}$$

$$\tau_{yz} = \frac{K_{III}}{(2\pi r)^{1/2}} \cos \frac{\theta}{2} \tag{8-b}$$

where K_{III} is the mode III stress intensity factor. In the simple case considered also by Griffith, the mode I stress-intensity factor is given by (Fig. 6):

$$K_I = \sigma\sqrt{\pi a} \tag{9}$$

Since the stress field in the crack tip vicinity is uniquely defined by the K_I factor, it is consistent to assume the instability to occur when K_I achieves its critical value K_{IC}:

$$\sigma \geq \frac{K_{IC}}{\sqrt{\pi a}} \tag{10}$$

Recalling that eqs. (5) and (10) refer to the same case, the LEFM relationship between critical stress-intensity factor and fracture energy is:

$$K_{IC} = \sqrt{E\, G_{IC}} \tag{11}$$

From the preceding formulas it is seen that the physical dimensions of the stress-intensity factor are rather unusual: $[K_I] = [F][L]^{-3/2}$. Such dimensions imply size effects in linear elastic fracture mechanics as well as, indirectly, in material strength. Equivalently, it is possible to assert that, while in the traditional Solid Mechanics we refer to energy dissipated in the unit volume (e.g., the equivalent Von Mises stress), in LEFM we refer to energy dissipated in the unit surface (e.g., the fracture energy G_{IC}). It is therefore evident that the co-existence of the two collapse mechanisms (plasticity and fracture) leads to peculiar size effects.

Only the simplest geometry has been considered up to now to introduce the subject: an infinite cracked plate subjected to uniform stress. In practice, however, the structural elements present finite size and the stress-intensity factor depends, besides on external loads and crack length, on structure dimensions. For example, a plate of finite width (Fig. 8) has the related stress-intensity factor provided by the following expression:

$$K_I = \sigma \sqrt{\pi a} \; f\left(\frac{a}{b}\right) \tag{12}$$

where the shape function f is listed in Broek (1986):

$$f\left(\frac{a}{b}\right) = \left(\sec \frac{\pi a}{2b}\right)^{1/2} \tag{13}$$

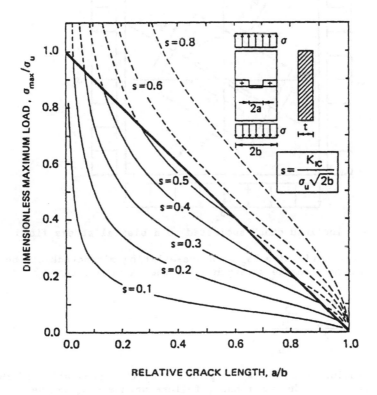

Fig. 8. Stress of crack propagation versus relative crack length (plate of finite width).

1.4   Crack and Structure Size Transitions

According to eqs. (5) and (10), the strength of a center cracked plate decreased by increasing the length 2a of the initial crack (Fig. 5). The strength tends to zero for a → ∞, whereas it asymptotically tends to infinity for a → 0. In fact, the theoretical strength of several materials, computed on the basis of the interatomic forces, is 10 to 100 times higher than the experimental one. It is reasonable to assume that this inconsistency could be attributed to the presence of inherent defects.

The theoretical model presents a single crack (Figs. 6 and 9) whereas real materials usually contain a greater number of flaws of different length and orientation. If the dominant crack is not sufficiently long and inclined with respect to the tensile direction, the stress at failure is achieved due to a natural crack which is more dangerous than the artificial one, Carpinteri et al. (1979).

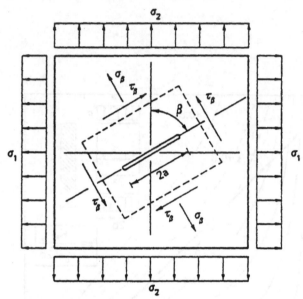

Fig. 9.   Inclined crack subjected to a biaxial stress field.

The characteristic size $2a_o$ of a pre-existing microcrack can be obtained from the tensile strength σ_u of the material, through eqs. (5) or (10):

$$a_o = \frac{E \, G_{IC}}{\pi \sigma_u^2} = \frac{K_{IC}^2}{\pi \sigma_u^2} \qquad (14)$$

Therefore, failure when $a > a_o$ is produced by the propagation of the dominant crack. On the other hand, failure when $a < a_o$ can be considered as caused by an inherent microcrack, while the crack of length 2a cannot be considered as dominant any more (Fig. 5).

It is possible to conclude that a crack size transition connects the two limit-cases of ultimate strength failure (small cracks) and LEFM crack propagation (large cracks), Carpinteri et al. (1979).

Only plates of infinite size have been considered up to now. If cracked bodies of finite size are taken into consideration, it is possible to demonstrate that a size-scale transition does exist analogous to the previous one. Comparatively small bodies are likely to undergo ultimate strength failure, whereas comparatively large bodies collapse due to brittle crack propagation, Carpinteri (1981, 1982).

Due to the different physical dimensions of tensile strength σ_u and fracture toughness K_{IC}, scale effects are always present in the usual fracture tests. This means that, for the usual size scale of the laboratory specimens, the ultimate strength collapse or the plastic collapse at the ligament tends to anticipate the real propagation of the initial crack, Carpinteri (1981, 1982).

Such a competition between collapses of a different nature can be easily proved by considering eqs. (12) and (13), which provide the stress-intensity factor for a cracked plate of finite width (Fig. 8). Eq. (12), in the condition of incipient crack propagation, particularizes as follows:

$$K_{IC} = \sigma_{max} \sqrt{\pi a} \; f\left(\frac{a}{b}\right) \tag{15}$$

If both sides of eq. (15) are divided by $\sigma_u \sqrt{2b}$, we obtain:

$$\frac{K_{IC}}{\sigma_u \sqrt{2b}} = s = \frac{\sigma_{max}}{\sigma_u} \sqrt{\frac{\pi a}{2b}} \; f\left(\frac{a}{b}\right) \tag{16}$$

where s is a dimensionless number able to describe the brittleness of the system and where both material properties and specimen size appear. Rearranging eq. (16) gives:

$$\frac{\sigma_{max}}{\sigma_u} = s \left(\frac{\cos \frac{\pi a}{2b}}{\frac{\pi a}{2b}}\right)^{1/2} \tag{17}$$

On the other hand, it is possible to consider, in an approximate sense, the non-dimensional load associated with the ultimate strength σ_u at the ligament (b-a):

$$\frac{\sigma_{max}}{\sigma_u} = 1 - \frac{a}{b} \tag{18}$$

Eqs. (17) and (18) are plotted in Fig. 8 as functions of the crack depth a/b. While the former provides a family of curves by varying the brittleness number s, the latter is represented by a unique curve (thick line). It is seen that ultimate strength collapse at the ligament precedes crack propagation for each initial crack depth when the brittleness number s is higher than the critical value $s_o = 0.54$.

For lower s numbers, ultimate strength collapse anticipates crack propagation only for initial crack depths external to a certain interval. This means that a real separation phenomenon occurs only with a relatively low fracture toughness, high tensile strength and/or large structure size. Not the single values of K_{IC}, σ_u and b, but only their function s - see eq. (16) - determines the nature of the collapse mechanism.

This discussion presents a simple, and approximate, illustration of the use of LEFM concepts to explain size effects on fracture. Extensive work has been done by a number of investigators to determine the limits of applicability of LEFM with respect to structure size, and also to consider the non-linear effects such as micro-cracking as an alternative, or modification, to LEFM. This body of work is presented in an earlier chapter of the RILEM report. Aspects of this as they pertain to mixed mode cracking and fracture will be presented in later sections.

2 Mixed-Mode Fracture Criteria

The various models which have been proposed to explain crack propagation in mixed-mode conditions may generally be classified as earlier and later mixed mode fracture theories.

2.1 Earlier Mixed-Mode Fracture Theories
These are the maximum circumferential stress; maximum strain energy release rate; minimum strain energy density; maximum circumferential strain; shear-friction; Mohr-Coulomb; and octahedral stress failure theories. In all of these the tie-in with fracture mechanics is the use of stresses expressed in terms of K_I and K_{II}.

For mixed-mode problems the stresses are conveniently expressed in polar (r, θ) coordinates as σ_r, σ_θ, $\tau_{r\theta}$. The origin of coordinates is taken at the crack tip and $\theta = 0$ corresponds to the crack line. Then, from LEFM, Broek (1986):

$$\sigma_r = \frac{1}{\sqrt{2\pi r}} \left[K_I \left(\frac{5}{4} \cos \frac{\theta}{2} - \frac{1}{4} \cos \frac{3\theta}{2} \right) \right.$$

$$\left. + K_{II} \left(-\frac{5}{4} \sin \frac{\theta}{2} + \frac{3}{4} \sin \frac{3\theta}{2} \right) \right] \tag{19}$$

$$\sigma_\theta = \frac{1}{\sqrt{2\pi r}} \left[K_I \left(\frac{3}{4} \cos \frac{\theta}{2} + \frac{1}{4} \cos \frac{3\theta}{2} \right) \right.$$

$$\left. + K_{II} \left(-\frac{3}{4} \sin \frac{\theta}{2} - \frac{3}{4} \sin \frac{3\theta}{2} \right) \right] \tag{20a}$$

$$= \frac{1}{\sqrt{2\pi r}} \cos \frac{\theta}{2} \left[K_I \cos^2 \frac{\theta}{2} - \frac{3}{2} K_{II} \sin \theta \right] \tag{20b}$$

$$\tau_{r\theta} = \frac{1}{2\sqrt{2\pi r}} \cos \frac{\theta}{2} \left[K_I \sin \theta + K_{II} (3 \cos \theta - 1) \right] \quad (21)$$

Maximum circumferential stress criterion. This criterion was proposed by Erdogan and Sih (1963) and is based on the assumption that the crack initiates from its tip in a direction normal to the maximum circumferential stress σ_θ. In Fig. 9 an inclined crack at angle ß to the principal stress direction is subjected to the stresses at infinity σ_β, τ_β. By using eqs. (19-21) and Mohr's circle analysis one obtains

$$K_I = \left[\frac{\sigma_1 + \sigma_2}{2} + \frac{\sigma_1 - \sigma_2}{2} \cos 2 \beta \right] \sqrt{\pi a} \quad (22a)$$

$$K_{II} = \left[\frac{\sigma_2 - \sigma_1}{2} \sin 2 \beta \right] \sqrt{\pi a} \quad (22b)$$

If m is the ratio σ_1/σ_2 of the principal stresses, then equations (22) can be rewritten as:

$$K_I = \sigma_2 \sqrt{\pi a} \left[m + (1 - m) \sin^2 \beta \right] \quad (23a)$$

$$K_{II} = \sigma_2 \sqrt{\pi a} (1 - m) \sin \beta \cos \beta \quad (23b)$$

Crack propagation occurs at an angle θ (c.f. Fig. 10) that maximizes σ_θ (i.e., σ_θ becomes a principal stress and $\tau_{r\theta} = 0$). Thus

$$\text{Tan} \frac{\theta}{2} = \frac{1}{4} \frac{K_I}{K_{II}} \pm \sqrt{\left(\frac{K_I}{K_{II}} \right)^2 + 8} \quad (24)$$

The crack propagation angle, θ and crack inclination angle, ß are related by:

$$[m + (1-m) \sin^2 \beta] \sin \theta + [\tfrac{1}{2} (1-m) \sin 2 \beta] (3 \cos \theta - 1) = 0 \quad (25)$$

Equation (25) can be transformed into:

$$2(1-m) \sin 2 \beta \left(\tan \frac{\theta}{2} \right)^2 - 2 [m + (1-m) \sin^2 \beta] \left(\tan \frac{\theta}{2} \right)$$
$$- (1-m) \sin 2 \beta = 0 \quad (26)$$

The solution of interest is reported in Fig. 10 for different ratios m.

If $m = 1$, it is always $\theta = 0$, since the stress field at infinity is uniform, and then the crack extension is collinear due to symmetry. If $m = 0$, there is a discontinuity for $\beta = 0$. In fact $\theta(\beta = 0, m = 0) = 0$ due to the symmetry, while

$$\lim_{\beta \to 0^+} \theta(\beta, m = 0) \simeq 70°$$

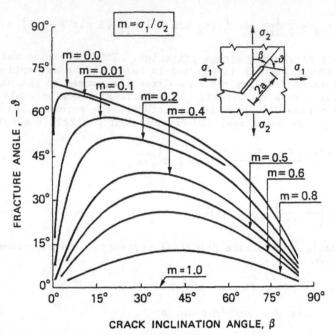

Fig. 10. Crack branching angle against crack inclination angle
(maximum stress criterion).

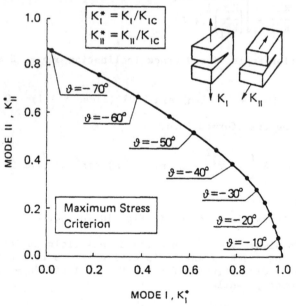

Fig. 11. Fracture locus in the plane of the stress-intensity
factors (maximum stress criterion).

If m is small but different from zero, the discontinuity disappears and is replaced by a rapid variation, represented by a very steep branch in Fig. 10. From a mathematical point of view, this is a typical case of <u>non-uniform convergence</u> of function $\theta(m,\beta)$ in $\beta = 0$ and for $m \to 0^+$.

While equation (24) determines the direction of maximum stress, an additional condition may be introduced as a criterion of instability:

$$\sqrt{2 \pi r} \ \sigma_\theta = K_{IC} \tag{27}$$

Introducing the dimensionless stress-intensity factors

$$K_I^* = K_I/K_{IC} \qquad K_{II}^* = K_{II}/K_{IC} \tag{28}$$

the fracture locus may be plotted in the K_I^* versus K_{II}^* plane. Solving:

$$K_I^* \sin \theta + K_{II}^* (3 \cos \theta - 1) = 0 \tag{29a}$$

$$K_I^* \cos^2 \frac{\theta}{2} - \frac{3}{2} K_{II}^* \sin \theta = \frac{1}{\cos \frac{\theta}{2}} \tag{29b}$$

for K_I^* and K_{II}^* by varying θ, all the points of the locus are defined. They are symmetric with respect to the K_I-axis and valid only in the half-plane $K_I^* \geq 0$, (Fig. 11). It was proved by Nuismer (1975) that the maximum stress criterion is totally equivalent to the maximum strain energy release rate criterion.

<u>Minimum strain energy density criterion</u>. The strain energy density function at the crack tip is of the form, Sih (1973); Sih and Macdonald (1974):

$$\left(\frac{dW}{dV}\right) = \frac{1}{r} (a_{11}K_I^2 + 2a_{12}K_IK_{II} + a_{22}K_{II}^2 + a_{33}K_{III}^2) \tag{30}$$

where the volume of the differential element is $dV = rdrd\theta$. The coefficients a_{ij} for plane strain are:

$$a_{11} = \frac{1}{16\pi G} [(3 - 4\nu - \cos \theta)(1 + \cos \theta)] \tag{31a}$$

$$a_{12} = \frac{1}{16\pi G} (2 \sin \theta) [\cos \theta - 1 + 2\nu] \tag{31b}$$

$$a_{22} = \frac{1}{16\pi G} [4(1 - \nu)(1 - \cos \theta) + (1 + \cos \theta)(3 \cos \theta - 1)] \tag{31c}$$

$$a_{33} = \frac{1}{4\pi G} \tag{31d}$$

where G is the tangential elastic modulus.
The function dW/dV thus possesses a 1/r singularity at the crack tip.

141

A strain energy density factor S can thus be defined as a function of θ:

$$\left(\frac{dW}{dV}\right) = \frac{S(\theta)}{r} \tag{32}$$

Sih (1973, 1974) proposed the following criteria:

(a) The crack initiation direction is assumed to correspond with the minimum strain energy density factor:

$$\frac{\partial S}{\partial \theta} = 0 \qquad \frac{\partial^2 S}{\partial \theta^2} > 0 \qquad \text{for } \theta = \theta_o \tag{33}$$

(b) The crack starts to propagate when S reaches a critical value, S_c at $\theta = \theta_o$.
For mode I crack extension, $K_{II} = K_{III} = 0$; $\theta_o = 0$ and therefore $a_{11} = (1 - 2\nu)/4\pi G$. The critical value of the strain energy density factor, S_c, can be related to the critical value of the stress-intensity factor, K_{IC}:

$$S_c = \frac{(1 - 2\nu)(1 + \nu)}{2\pi E} K_{IC}^2 \qquad \text{for plane strain.} \tag{34}$$

In the case of mixed-mode in-plane loading, $K_I \neq 0$, $K_{II} \neq 0$ and $K_{III} = 0$. By the application of equations (30) and (31), an expression for finding the stationary values of S is obtained:

$$\frac{\partial S}{\partial \theta} = K_I^2 \frac{1}{8G} \sin\theta(\cos\theta - 1 + 2\nu) + 2K_I K_{II} \frac{1}{8G} [2\cos^2\theta$$

$$+ (2\nu - 1)\cos\theta - 1] + K_{II}^2 \frac{1}{8G} \sin\theta(-3\cos\theta + 1 - 2\nu)$$

$$= 0 \tag{35}$$

This results in a fourth order equation:

$$2(1-\nu)K_I K_{II}\left(\tan\frac{\theta}{2}\right)^4 + (3 K_{II}^2 - K_I^2)\left(\tan\frac{\theta}{2}\right)^3 + [(2\nu - 1)(K_I^2 - K_{II}^2)$$

$$- 6 K_I K_{II}]\left(\tan\frac{\theta}{2}\right)^2 + [K_I^2 - 3 K_{II}^2]\left(\tan\frac{\theta}{2}\right)$$

$$+ [(2\nu - 1)(K_I^2 - K_{II}^2) + 2 K_I K_{II} \nu] = 0 \tag{36}$$

The two factors, K_I and K_{II}, are functions of the loading ratio m, and crack inclination angle ß.
Substituting equations (23) into (36), it is possible to obtain the crack inclination angle, ß, as a function of θ. The results are displayed graphically in Fig. 12 for $\nu = 0.3$.
The crack growth conditions are:

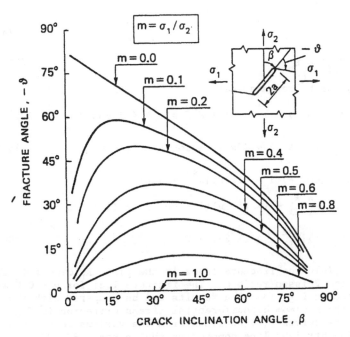

Fig. 12. Crack branching angle against crack inclination angle
(minimum strain energy density criterion).

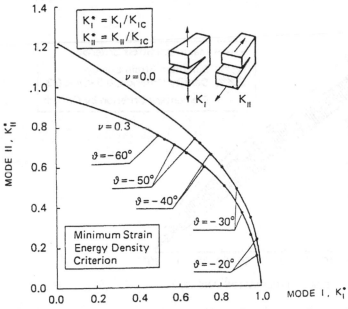

Fig. 13. Fracture locus in the plane of the stress-intensity factors
(minimum strain energy density criterion).

$$\frac{\partial S}{\partial \theta} = 0 \qquad (37a)$$

$$S(K_I, K_{II}) = S_c \qquad (37b)$$

Equations (37) can be restated as:

$$K_I^{*2} \sin \theta (\cos \theta - 1 + 2\nu) + K_{II}^{*2} \sin \theta (-3 \cos \theta + 1 - 2\nu)$$

$$+ 2 K_I^* K_{II}^* [2 \cos^2 \theta + (2\nu - 1) \cos \theta - 1] = 0 \qquad (38a)$$

$$[(3 - 4\nu - \cos \theta)(1 + \cos \theta)]K_I^{*2} + 4 \sin \theta [\cos \theta$$

$$- (1 - 2\nu)]K_I^* K_{II}^* + [4 (1 - \nu)(1 - \cos \theta) + (1 + \cos \theta)$$

$$(3 \cos \theta - 1)]K_{II}^{*2} - 4 (1 - 2\nu) = 0 \qquad (38b)$$

Equations (38) yield the fracture locus in the plane K_I^* versus K_{II}^* for each value of the Poisson ratio ν. The fracture loci for $\nu = 0.0$ and 0.3 are reported in Fig. 13. The results can be compared with those obtained from the maximum circumferential stress criterion in Fig. 11.

In Fig. 14 the curves $\theta(\beta)$ obtained from the various theories are summarized. The straight line connecting the points - $\theta = 90°$ and $\beta = 90°$ corresponds to the simple assumption of Griffith that the crack propagates orthogonally to the direction of the uniaxial applied stress.

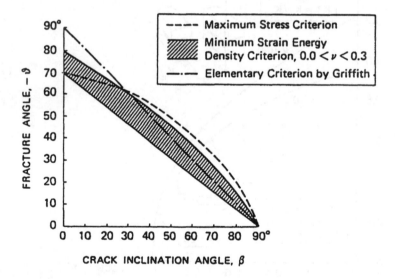

Fig. 14. Crack branching angle against crack inclination angle: comparison between the various theories.

The Mohr's circle in Fig. 15 represents all the pairs of values of normal stress σ_β and shear stress τ_β obtained by varying the crack orientation, when the maximum tensile stress is equal to the ultimate strength σ_u. On the other hand, the mixed mode crack propagation locus in Fig. 15 shrinks by increasing the length of the dominant crack. When a → ∞, the intersection locus tends to coincide with the fracture mechanics locus and the allowable stresses tend to zero. When a = a_o, the intersection locus coincides with the critical Mohr's circle (Fig. 15) and the body becomes macroscopically uncracked, i.e., the dominant crack in this case presents the size or, generally, the failure potentiality of the inherent microcracks. Then, for a < a_o, the crack cannot be considered as dominant and is totally negligible independently of its orientation.

When a_o < a < ∞, the macrocrack propagation occurs only for crack inclination angles $\beta \geq \beta_o(a)$, β_o being the angle that the straight line connecting the origin with the cuspidal point of the intersection locus forms with the positive τ_β semi-axis. The limit-angle β_o versus the ratio a/a_o is presented in Fig. 16. When a → ∞, macrocracks can propagate for any orientation (β_o → 0). When a = a_o and according to the Strain Energy Density Criterion, β_o = 90° and the crack can propagate only when it is orthogonal to the applied stress.

Maximum circumferential strain criterion. Crack propagation occurs when ϵ_θ reaches a limiting value $\epsilon_{\theta m}$, Maiti and Smith (1984).

$$\epsilon_\theta = \frac{1}{E\sqrt{2\pi r}}\left\{K_I\left[\cos\frac{\theta}{2}\left(\frac{3}{4} - \frac{5}{4}\nu\right) + \frac{1}{4}\cos\frac{3\theta}{2}(1+\nu)\right]\right.$$

$$\left. + K_{II}\left[\sin\frac{\theta}{2}\left(-\frac{3}{4} + \frac{5}{4}\nu\right) - \frac{3}{4}\sin\frac{3\theta}{2}(1+\nu)\right]\right\} \qquad (39)$$

where ν is Poisson's ratio for concrete - about 0.15 to 0.20. The critical value, $\epsilon_{\theta m}$ may be obtained by considering $\theta = 0$, $\sigma_r = 0$ and is

$$\epsilon_{\theta m} = \frac{K_{Ic}}{E\sqrt{2\pi r}}(1-\nu) \qquad (40)$$

Shear-friction. Melin (1986) showed that for mode II crack growth, the crack will be extended without directional change, but if mode I starts to dominate, the crack will be extended via a kink. A crack of length 2a, subjected to different values of pressure p and shear stress τ (Fig. 17a), was studied. The crack was assumed to form at the angle θ shown.

The maximum value of K_{II} was determined to be associated with the following values of θ:

$$\theta = \theta_d = \begin{bmatrix} 0 & \text{if } p/\tau \leq 0 \\ \frac{1}{2}\sin^{-1}(p/\tau) & \text{if } 0 < p/\tau \leq \mu/(1+\mu^2)^{\frac{1}{2}} \\ \frac{1}{2}\tan^{-1}\mu & \text{if } \mu/(1+\mu^2)^{\frac{1}{2}} < p/\tau < (1+\mu^2)^{\frac{1}{2}}/\mu \end{bmatrix} \qquad (41)$$

where μ is the coefficient of friction along the crack's surfaces. A crack originally situated in this direction will grow in mode II without change of direction with a stress intensity factor

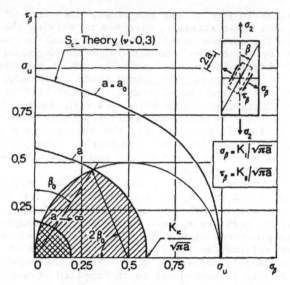

Fig. 15. Intersection locus in the Mohr's plane.

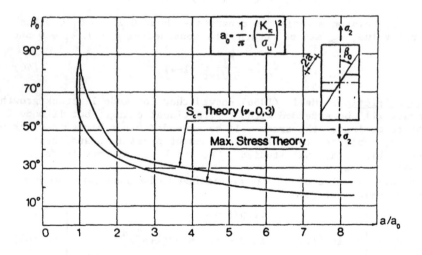

Fig. 16. Limit-angle β_0 against dimensionless crack length.

$$K_{IImax}/(\pi a)^{\frac{1}{2}} = \begin{bmatrix} \tau & \text{if } p/\tau \leq 0 \\ (\tau^2 - p^2)^{\frac{1}{2}} & \text{if } 0 < p/\tau \leq \mu/(1+\mu^2)^{\frac{1}{2}} \\ -\mu p + \tau(1+\mu^2)^{\frac{1}{2}} & \text{if } \mu/(1+\mu^2)^{\frac{1}{2}} < p/\tau \\ & < (1+\mu^2)^{\frac{1}{2}}\mu \end{bmatrix} \quad (42)$$

provided that K_{IImax} reaches the critical value for growth, K_{IIc}, and that mode I growth via a kink, for some reason, is prevented.

Melin assumed a very small kink at a different angle from the original crack (Fig. 17b) and calculated the stress intensity factors at the tip of the kink. He found that K_I for the kink is maximum at an angle of about 70 degrees from the original crack (Fig. 17c). He also found that if the crack continues to grow in mode I, it will follow a smooth path toward the direction perpendicular to the largest principal stress and the crack will be open so that no frictional sliding is involved. A typical contour of the separated surface of the main crack and the kink is shown in Fig. 17c.

A criterion for deciding whether the crack will grow in mode I by formation and extension of a kink or grow in mode II in the original crack direction might be established by comparing the stress intensity factors K_{Imax} for the small kink with K_{IImax} for the main crack. If K_{Ic} and K_{IIc} denote the critical stress intensity factors for mode I and mode II respectively, mode II crack growth will be preferred to mode I if

$$k = \frac{K_{IImax}}{K_{Imax}} > \frac{K_{IIc}}{K_{Ic}} = k_c \quad (43)$$

For the case of $\mu = 0.4$ it was concluded that the crack extension will be in mode I if $p/\tau < 2.2$, will continue in mode II if $2.2 \leq p/\tau \leq 2.4$ and no growth can occur if $p/\tau > 2.7$.

Other failure theories. Other failure theories could be used in the same way as the ones just described. It is possible to consider the Mohr-Coulomb theory and the octahedral stress theory.

Mohr-Coulomb:

$$\tau_{r\theta} \geq C_1 + C_2\, \sigma_\theta \quad (44)$$

where C_1, C_2 may be determined from conditions associated with K_{Ic}, K_{IIc}.

Octahedral Stress:

$$\text{a.} \quad \tau_o = \frac{\sqrt{2}}{3}\left\{\left[\frac{\sigma_r + \sigma_\theta}{2}\right]^2 + 3\left[\left(\frac{\sigma_r - \sigma_\theta}{2}\right) + \tau_{r\theta}^2\right]\right\}^{\frac{1}{2}} \quad (45)$$

and propagation occurs when $\tau_o = \tau_{om}$; or,

$$\text{b.} \quad \tau_o = C_3 + C_4 \sigma_o \quad (46)$$

147

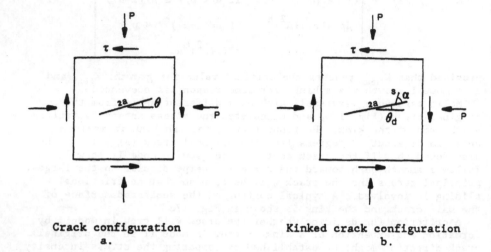

Crack configuration
a.

Kinked crack configuration
b.

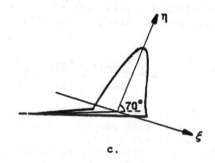

c.

Fig. 17.   Formation of a kink.

where $\sigma_o = \frac{1}{3}(\sigma_r + \sigma_\theta)$ and C_3, C_4 may be determined from conditions associated with K_{Ic} and K_{IIc}.

2.2 Later Mixed-Mode Fracture Theories
Energy Balance. The total energy release rate based on LEFM may be

(see Broek (1986))

$$G_t = G_I + G_{II} \tag{47}$$

where, for concrete and mode I (c.f. eq. 11)

$$G_I = \frac{K_I^2}{E} \tag{48a}$$

and for mode II

$$G_{II} = \frac{K_{II}^2}{E} \tag{48b}$$

In this the effect of Poisson's ratio is ignored.

Crack extension occurs along a direction that maximizes G_t and propagation occurs when $G_t = G_c$, a critical value. Or, as suggested by Jenq and Shah (1988)

$$K_I^2 + K_{II}^2 = K_c^2 \tag{49}$$

where K_c is defined to be the total stress intensity factor and K_I, K_{II} are the stress intensity factors of the kink. A variation on this is described by Di Leonardo (1979) as

$$K_I^2 + q^2 K_{II}^2 = K_{Ic}^2 \tag{50}$$

and, alternatively, Zhou (1987)

$$\lambda_{12} K_I + K_{II} = K_{IIc} \tag{51}$$

A more general approach requiring four parameters would be

$$\left[\frac{K_I}{K_{Ic}}\right]^\alpha + \left[\frac{K_{II}}{K_{IIc}}\right]^\beta = 1 \tag{52}$$

where α, β are material exponents - see for instance Broek (1986). Alternative formulations exist in terms of energy release rates if K_I, K_{II} are replaced by G_I, G_{II} by eqs. (48).

149

<u>Jenq/Shah</u>. This model uses two fracture mechanics parameters, $K_{IC}(= K_{Ic}^s)$ and $CTOD_c$, Jenq and Shah (1985, 1988). The first of these is determined using an effective crack length which accounts for the presence of a process zone in an approximate manner. The second parameter is the critical crack-tip-opening displacement. This can be related to the crack-mouth-opening displacement which can be readily measured.

For mixed-mode crack propagation, the total stress intensity factor, K_c (cf eq. 49) is used as a fracture parameter and a crack-tip displacement, CTD is defined to be the vector sum of the sliding and opening mode displacements as shown in Fig. 18. The fracture criteria were proposed to be $K_c = K_{IC}^s$ and $CTD = CTD_c = CTOC_c$.

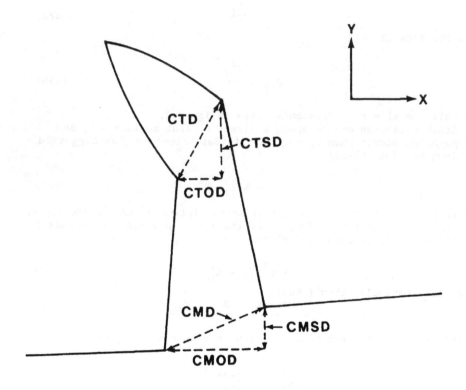

Fig. 18. Compositions of crack tip displacement and crack mouth displacement.

<u>Kokini</u>. This method defines an "effective stress intensity factor" as

$$K_f = \lim_{r \to 0} \sqrt{2\pi r} \; (\sigma_\theta)_{max} \qquad (53)$$

where $(\sigma_\theta)_{max}$ is obtained from eq. (20) as a principal stress. The propagation criterion is that $K_f = K_{1c}$.

2.3 Mixed-Mode Crack Initiation In Concrete

The crack models and the fracture propagation criteria described in the previous sections are directly applicable to ideal homogeneous and linear elastic materials. They may not be appropriate to simulate a heterogeneous and non-linear material like concrete. On the other hand, by increasing the size-scale of a concrete element the influence of heterogeneity disappears and the body can be considered macroscopically homogeneous. Moreover, by increasing the size-scale of a cracked concrete element, the influence of the non-linear softening material behavior vanishes, the cohesive crack tip forces disappear and the crack propagation is governed only by the linear elastic stress-singularity in the crack tip region, Peterson (1981); Carpinteri (1985); Carpinteri et al. (1986).

The preceding observations allow us to consider crack size and structure size transitions even for concrete-like materials. That is, we can assume LEFM as a valid crack branching criterion for large cracks in concrete and, more generally, for large concrete structures. For mode I crack propagation, this has already been demonstrated, Bazant (1984); Carpinteri (1981, 1982, 1985); Carpinteri et al. (1986); Ingraffea (1985); Peterson (1981); Wecharatana and Shah (1983). For mixed-mode crack propagation, this is a logical consequence, since the mode II stress-singularity power is still 1/2 as for mode I, Carpinteri (1987). On the other hand, for small pre-cracked concrete elements, an ultimate tensile strength collapse does precede the brittle crack branching (or, equivalently, we can say that the propagation of a natural crack does precede that of the dominant crack).

3 Test Specimen Geometries

3.1 Overview

Many different testing geometries have been used in an attempt to induce both in-plane shear and mode II crack initiation and extension. A number of these are shown in Figs. 19, 20, 25, 27-29, 33, 37-42, 44 and 45. Many analyses and experimental results obtained using these various geometries have appeared in the literature, Arrea and Ingraffea (1981); Ballatore et al. (1989); Barr and Hughes (1988, 1989); Barr et al. (1987); Bazant and Pfeiffer (1985a, 1985b, 1986); Bocca et al. (1989); Di Leonardo (1979); Ingraffea and Panthaki (1985); Izumi et al. (1986); Jenq and Shah (1988); Kokini et al. (1987); Rots and de Borst (1987); Swartz et al. (1988a, 1988b); Swartz and Taha (1989); Taha (1988); van Mier and Nooru-Mohamed (1988); Zhou (1987). This report is limited to a discussion of the following geometries: three that were proposed to subcommittee C of RILEM 89-FMT, (1) three point bending beam with off-center notch, (2) four point shear specimen, (3) quadratic double-edge-notched specimen; and two others, (4) Arcan specimen, Arcan et al. (1978); Arcan and Banks-Sills (1982); Banks-Sills and Bortman (1986), (5) semi-circular bend specimen, Chong and Kuruppu (1984); Chong et al. (1989). These will be presented and discussed in the following sections. Whereas the

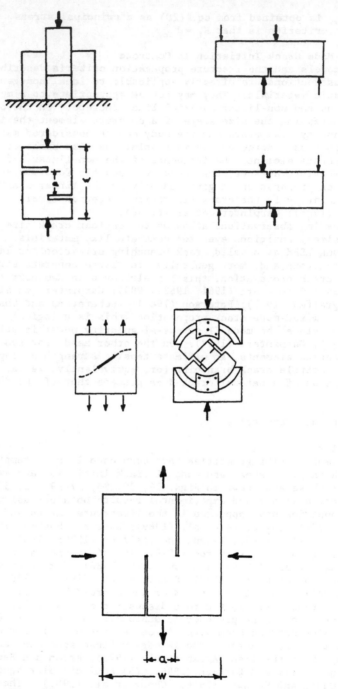

Fig. 19. Various geometries for mixed-mode tests.

first two testing geometries aim at evaluating mode II and/or mixed-mode fracture energy and/or fracture toughness, and at analysing the phenomenon of crack propagation, the third is utilized to study the constitutive laws of a fracture zone, i.e. normal and tangential stress vs. crack opening and sliding displacement. The two objects are connected. On the other hand, whereas testing methods (1) and (2) imply the existence of a crack tip evolving in the specimen, testing method (3) does not. The fourth geometry was proposed for testing composites and metals but may be adapted to concrete to study crack propagation. The fifth geometry, originally proposed to be used with rock cores, could be used equally well with concrete cores.

3.2 Three-Point-Bending Beam with Off-Center Notch

Jenq and Shah (1988) conducted mixed-mode experiments using beams which had notches at different locations along the span (Fig. 20). Depending on the geometries of the specimens, the final crack may not occur at the location of the notch. Various crack initiation criteria were discussed and compared with the experimental results. Agreement between the experimental results and the theoretical predictions on the crack initiation angles, final failure locations and peak loads of the tested specimens was judged to be quite satisfactory.

It was found that the crack initiation angles are very difficult to measure due to the tortuosity of crack paths. However, it can be shown that the crack will propagate in a straight line along the predicted initiation angle as shown in Fig. 21. The initiation angle can then be assumed as the final failure angle measured from notch-tip to final failure point. Fig. 22 gives the theoretical predictions of crack initiation angles along with the experimental results of all specimens tested. Despite the large scattering observed from the experimentally measured final crack angles, the theoretical predictions of crack initiation angles seem to be reasonable.

The peak load was underestimated by the conventional maximum energy release rate criterion for large beams and overestimated for small beams due to the size effect. Furthermore, using the conventional maximum energy release rate criterion, the failure location will always occur at the location of the notch and the predicted failure load will approach infinity when the off-set ratio is close to unity (Fig. 23) or the notch-depth ratio approaches zero (Fig. 24). This prediction is in contradiction with the experimental results. It shows that the conventional single fracture parameter approach cannot be used to characterize the general fracture behavior of concrete.

In the proposed model, the tractions acting along the crack surface were assumed to be negligible. This assumption may be reasonable for the present testing configuration in which the K_{II}/K_I ratios are relatively small. However, for higher K_{II}/K_I ratios, it was speculated that the tractions that occurred due to aggregate interlock effect may not be negligible and have to be considered in the analysis.

A similar three point bending configuration with two notches (Fig. 25) was investigated by Swartz et al. (1988a). Six notch locations, x, were used with six notch depths for each location for a total of 36 tests. The fracture energy was calculated using the total energy absorbed up to peak load divided by the inclined surface area. The latter was obtained from measurement of the approximate crack angle as

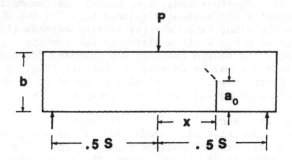

Fig. 20.   Three point bend beam with off center notch.

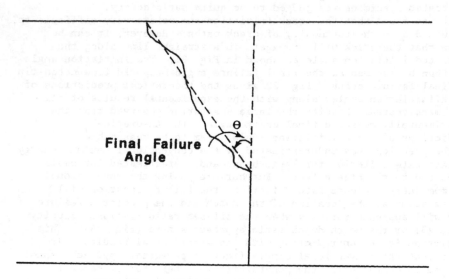

**Final Failure
Angle**

θ

Fig. 21.   Definition of final failure angle.

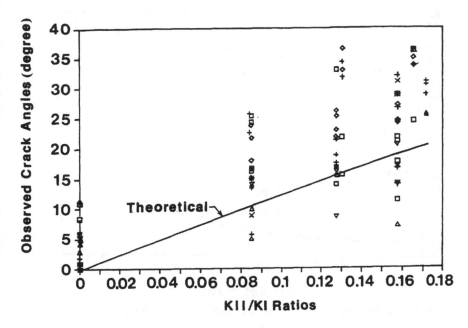

Fig. 22. Comparisons of theoretical predictions and experimental measures of final failure angle-all tested specimens.

in Fig. 21. Using this approach, additional energy absorbed beyond the peak load is neglected and also any effects of snapback are neglected. The results are presented for each notch location in Fig. 26 in which it is seen that higher values of fracture energy are obtained for locations near the support and for lower notch-depth ratios. This implies a strong influence of aggregate interlock and surface friction on the fracture behavior. A similar variation with notch location is shown when energy change is considered (J_c).

3.3 Four-Point-Shear Specimen
This specimen was proposed by Iosipescu (1967) and has been used in a variety of modifications. In Fig. 27 is shown the setup of Kumosa and Hull (1987) which fairly accurately follows the original concept. Three salient points to be noted are (1) the geometry provides constant shear and almost zero moment at the test section; (2) the test section is doubly-notched (the original geometry had a groove around the entire test region); (3) the test fixture is designed to prevent rotation of the two halves of the test specimen.

Beams with approximately this type of geometry have been tested by a number of investigators, Arrea and Ingraffea (1981); Ballatore et al. (1989); Bazant and Pfeiffer (1985a, 1985b, 1986); Bocca et al. (1989); Ingraffea and Panthaki (1985); Swartz et al. (1988); Swartz and Taha (1989); Taha (1988). Symmetrically-notched beams as shown in Fig. 28 of concrete and mortar, loaded near the notches by concentrated forces that produce a concentrated shear force zone were tested

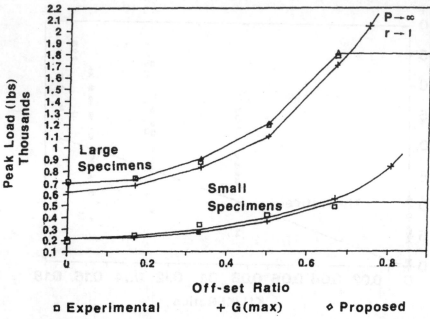

Fig. 23. Comparision of experimental peak load and theoretical predictions using two parameter fracture model and maximum energy release rate criterion for different offset ratios — C series.

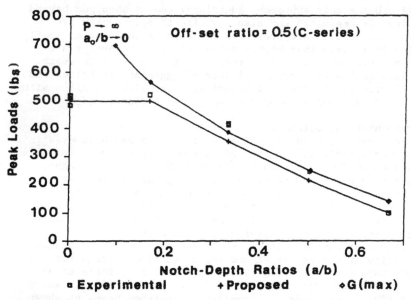

Fig. 24. Comparison of experimental peak load and theoretical predictions using two parameter fracture model and maximum energy release rate for different notch-depth ratios — C series.

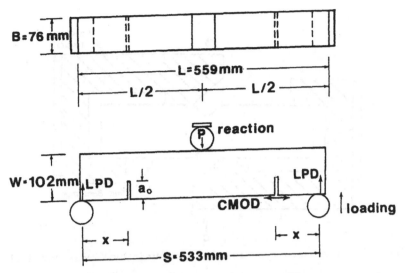

Fig. 25. Three point bend beam with two off-center notches.

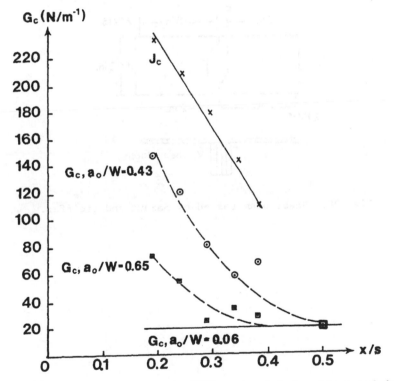

Fig. 26. Fracture energy at different notch locations and depth.

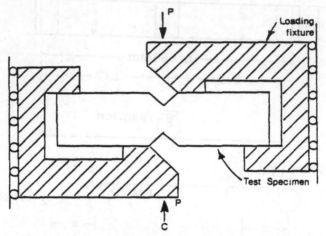

Fig. 27. Iosipescu-type specimen and used by Kumosa and Hull.

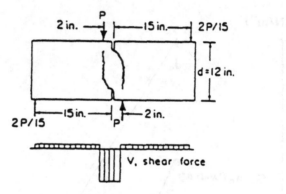

Fig. 28. Shear beam tested by Bazant and Pfeiffer.

to failure by Bazant and Pfeiffer (1985a, 1985b). They achieved the following conclusions.

1.   Shear fracture (i.e., mode II fracture) of concrete exists.

2.   Like the tensile (mode I) fracture, the shear (mode II) fracture follows the size effect law of blunt fracture. This implies that a large fracture process zone must exist at the fracture front, and that nonlinear fracture mechanics should be used, except possibly for extremely large structures.

3.   The maximum aggregate size D_{max} appears acceptable as a characteristic length for the size effect law. This further implies that the size of the fracture process zone at the maximum load is approximately a certain fixed multiple of the maximum aggregate size.

4.   The R-curve for the shear fracture may be obtained from the size effect law as the envelope of all fracture equilibrium curves for geometrically similar specimens of various sizes.

5.   The shear (mode II) fracture energy for the present type of test appears to be about 25 times larger than the tensile (mode I) fracture energy. This large difference may probably be explained by the fact that shear fracture energy includes not only the energy to create inclined tensile microcracks in the fracture process zone, but also the energy required to break the shear resistance due to interlock of aggregate and other asperities on rough crack surfaces behind the crack front.

6.   The direction normal to the maximum principal stress cannot be considered, in general, as the direction of crack propagation in concrete. Rather, fracture seems to propagate in the direction for which the energy release rate from the entire structure is maximized.

7.   The test specimen in Fig. 28 does not undergo perfectly antisymmetric deformation. Nevertheless, the symmetric deformation component has apparently only a small disturbing effect because, in the sense of linear elastic fracture mechanics, the Mode I stress intensity factor is not positive. According to both linear and nonlinear finite element analysis, the maximum principal tensile stresses occur at the notches. So the cracks must originate also at the notches and propagate inward. This means that the fracture mode is essentially shear and that the effect of the transverse tensile strain at the middle of the ligament cannot be significant.

The behavior emphasized by Bazant and Pfeiffer (1985a, 1985b) is in contrast to that observed in similar specimens tested by Arrea and Ingraffea (1981). Ingraffea and Panthaki (1985) showed by analysis - both classical elasticity and finite elements - that the specimens tested by Bazant and Pfeiffer (1985a, 1985b) fractured in a manner quite similar to that which occurs in a splitting tensile strength test. The following conclusions were drawn.

1.   Limited tensile cracking from the notch tips, although not reported by Bazant and Pfeiffer (1985a, 1985b) is likely to have occurred and should be observable using a crack detection technique more sensitive than the naked eye.

2.   By moving the principal loads closer together than in the Arrea and Ingraffea (1981) tests, the intensity of the shear stress in the region between notch tips is decreased rather than increased. The shear stress distribution, magnitude and

trajectory, in this region cannot be predicted using elementary beam theory.

3. The major principal stress in the region between the notches is tensile, and the direction is roughly constant at about 10 degrees above the horizontal. This means that the plane of principal shear is far from being vertical.

4. The stress state between the notch tips, despite the existence of the notches and the minor loads, is startlingly similar to that in the central region of a cube or cylinder loaded by diametrically opposed point loads, the Brazilian test.

5. By using the solution for the principal tension at the center of hypothetical Brazilian specimen inscribed within the beam specimen, one can predict the peak loads for the beam specimens with very good to good accuracy.

6. Based on the above conclusions, it is apparent that the beams suffered tensile, not shear, fracture, with a crack nucleating in the central region of the beam, roughly vertical, and propagating towards the notch tips.

Usually, shear failures may be imputed to tensile stress (Fig. 29a). Even the trajectory of a brittle crack always follows the direction normal to the local principal tensile stress (Fig. 29b), according to the maximum circumferential stress criterion described in section 3.2.1. Arrea and Ingraffea (1981) argued about a transition between two different failure modes by varying the distance c of the central supports. When the distance is large, the crack branching is favored and a crack trajectory develops from the crack tip to the opposite force (Fig. 29b). On the contrary, when the distance is small, the ultimate tensile strength is exceeded at the center of the specimen (Fig. 29c) and a nearly vertical crack nucleates far from the crack tips and propagates upwards and downwards as in the Brazilian Splitting Test. In addition, when the distance is small, the St. Venant's Principle is not valid between the central supports and the elementary beam theory fails completely (Fig. 29c).

It is clear now that, for the Ingraffea's test, there are two potential failure mechanisms in competition, by varying the distance c (Fig. 29b and c) as well as by varying the size-scale b of the specimen. Mixed mode crack branching is favored by increasing the distance c or the size-scale b (the ratio c/b being constant). On the contrary, ultimate tensile failure at the center is favored by decreasing the distance c or the size-scale b. This is the reason for which Ingraffea and Panthaki (1985) cannot predict the experimental results obtained by Bazant and Pfeiffer (1985a, 1985b) with large specimens, using only the simple Brazilian test formula.

In Ingraffea's test, stress intensification is produced in both the crack tip regions and the stress-intensity factors for mode I and mode II can be expressed respectively as (Fig. 29b and c):

$$K_I = \frac{P}{t\,b^{1/2}}\, f_I\left(\frac{1}{b}\,,\,\frac{a}{b}\,,\,\frac{c}{b}\right) \tag{54a}$$

$$K_{II} = \frac{P}{t\,b^{1/2}}\, f_{II}\left(\frac{1}{b}\,,\,\frac{a}{b}\,,\,\frac{c}{b}\right) \tag{54b}$$

160

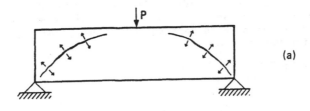

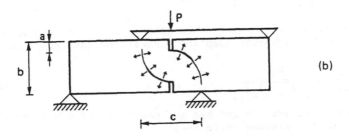

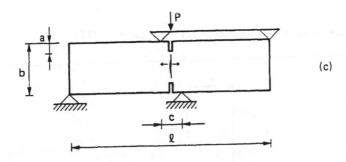

Fig. 29. The shear failures may be always imputed to tensile stress: (a) classical shear fracture in a concrete plate; (b) mixed-mode crack propagation; (c) tensile splitting in the center of the plate.

161

where f_I and f_{II} are the respective shape functions, which can be determined numerically, and t is the specimen thickness.

When the distance c tends to zero, the Mode I stress intensification vanishes and:

$$\lim_{\frac{c}{b} \to 0} K_I = 0 \tag{55}$$

Most relevant fracture criteria can be expressed in the approximate form, DiLeonardo (1979) of eq. (50) as:

$$K_I^2 + q^2 K_{II}^2 = K_{IC}^2 = G_{IC} E = G_{IF} E \tag{56}$$

where q is a measure of the influence of Mode II on crack propagation (it depends on the criterion adopted). Recalling eqs. (54), eq. (56) becomes:

$$\frac{P_{max}}{t\, b^{1/2}} = \frac{K_{IC}}{\sqrt{f_I^2 + q^2 f_{II}^2}} \tag{57}$$

If the geometric ratios l/b and a/b are constant and c/b is varying, eq. (57) can be transformed as follows:

$$\frac{P_{max}}{\sigma_u t b} = \frac{s}{F(c/b)} \tag{58}$$

where s is the brittleness number defined in eq. (16).

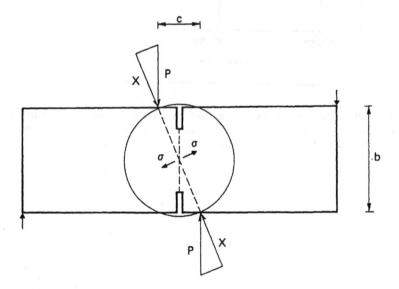

Fig. 30. Diametrically-loaded disk ideally inscribed in the plate.

On the other hand, according to the scheme in Fig. 30, we can ideally inscribe a diametrically loaded disk in the beam specimen and obtain the central tensile stress σ by the simple Brazilian Test formula:

$$\sigma = \frac{2\ Pb}{\pi t \left(b^2 + c^2\right)} \tag{59}$$

The maximum load due to the achievement of the tensile strength in the center of the specimen is given in dimensionless form:

$$\frac{P_{max}}{\sigma_u tb} = \frac{\pi}{2} \left[1 + \left(\frac{c}{b}\right)^2\right] \tag{60}$$

Eqs. (58) and (60) are in competition by varying the value of the brittleness number s or by varying the ratio c/b. If the latter is fixed, there is a size-scale transition which is described in Fig. 31. For small s values, i.e. for large sizes, LEFM crack branching precedes ultimate tensile strength failure and eq. (60) overestimates the maximum load [31]. Vice versa, for large s values, i.e. for small sizes, ultimate tensile strength failure precedes LEFM crack branching and eq. (58) overestimates the maximum load, which can be given by eq. (60) with a very good accuracy.

The experimental results by Bazant and Pfeiffer (1985a, 1985b) are reported in Fig. 32, where the maximum load divided by the beam area is plotted against the inverse root $b^{-1/2}$. The experimental points describe the transition between an inclined straight line passing through the origin and a horizontal asymptote. The former represents the LEFM instability, whereas the latter the ultimate tensile strength instability. It is important to realize that, with a mixed-mode crack configuration, larger (specimen and crack) sizes are needed than with a mode I crack configuration, in order to utilize LEFM correctly. For example, mortar specimens ~60 cm deep are necessary at least to apply mixed-mode LEFM in Fig. 32, whereas specimens 15 cm deep are usually sufficient for mode I. This considerable result is implicitly present also in Fig. 16.

Swartz et al. (1988b) presented experimental and analytical results and fracture parameters for beams loaded in four-point shear with a single edge, starter notch located in a region of high shear and low bending moment. The following conclusions were made.

1. If linear elastic fracture mechanics is assumed, the finite element analysis indicates that crack propagation starts in mode II at an angle of approximately ± 70°. However, in no case did this occur experimentally and the crack direction changed almost immediately after slow growth was initiated and became about 30° to 40°. In this mode the F.E. program indicates propagation in a mode I condition. It is also noted that the crack path is far from smooth and is likely associated with mixed-mode behavior (rather than mode I) due to aggregate interlock and friction.

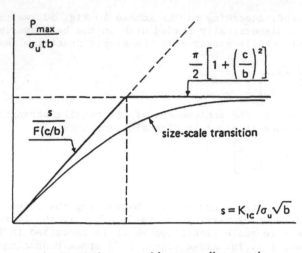

Fig. 31. Size-scale transition between LEFM and ultimate tensile strength.

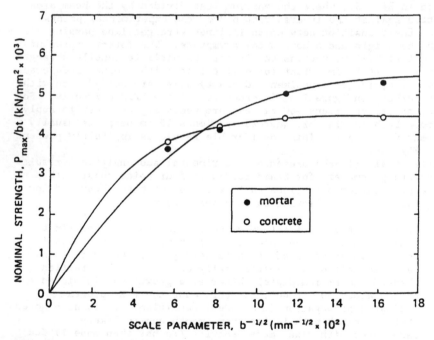

Fig. 32. Non-similarity in the mixed-mode fracture behavior of concrete: experimental results of Bazant and Pfeiffer.

2. The influences of aggregate interlock and friction forces must be great. At the end of the fracture process the crack surfaces are subjected to compressive stresses.
3. While the results of the energy estimates show similar variations with the notch depth a_0/W, as compared to analytical results the values are much different. These results become independent of notch depth at $a_0/W = 0.5$.
4. No matter which approach is used to determine K_{IIc} for concrete, the resulting value is much greater than K_{Ic}.
5. The presence of a possible process zone is not accounted for. This effect may be great for small beams and deserves further study. If the process zone is large then LEFM is not valid.
6. The fracture energy values are about 8-10 times those obtained for mode I tests on the same concrete using the same beam size and similar procedures. This is similar to that reported by Bazant and Pfeiffer (1985a, 1985b) in which they reported a 30 times difference.
7. This type of beam test does not at present appear to yield definitive estimates of either K_{IIc} or even mode II crack propagation. It may be possible to use this approach in conjunction with an appropriate analytical model in which aggregate interlock and friction are properly considered.

A second study was conducted by Swartz and Taha in which double-notched beams were used and, in some cases, with axial compression applied in addition to the shearing loads, Taha (1988); Swartz and Taha (1989). Their test setup is presented in Fig. 33 in which the axial compression system is also shown. This frame was always present, even if no axial load was applied and thus the beam boundary conditions were similar to the Iosipescu setup, c.f. Fig. 27.

They concluded that crack propagation in mixed mode appears to be due to mode I deformation even though mode II deformation and shear-friction are present. It was felt that a substantial amount of fracture energy is due to shear-friction which is not a fracture mechanics parameter in the usual sense.

Further conclusions for each type of test were as follows.

Beams Without Axial Force

1. The initial crack angle predicted by the maximum stress criterion of 70° (for pure mode II deformation) was in good agreement with the average of the experiments of 64°. This implies crack initiation in mode II.
2. The results from CRACKER, Swenson (1986) agreed with the experiments for crack path as shown in Fig. 34. The program indicates that the crack propagates in a direction associated with the local maximum principal stress and further that this is associated primarily with mode I deformation.
3. The CRACKER and ANSYS analyses both showed that neither the maximum tensile stress nor the shear stress exceeded the concrete strength along a section connecting the notch tips. Thus, these beams did not fail along that line.

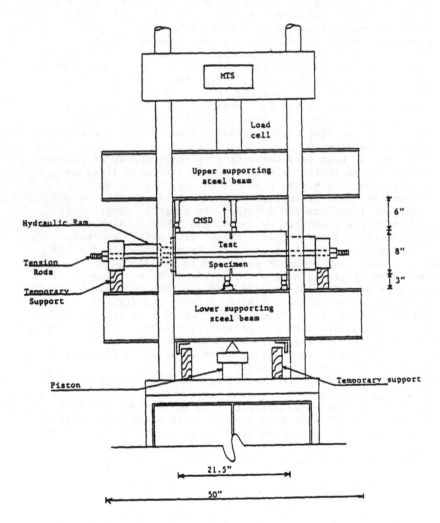

Fig. 33.   Overall test setup of Taha.

Beams With Axial Force

4. Application of the axial compression did not preclude
    development of tensile stresses at the crack tip and cracks
    initiated at an angle of 50° (compared to 64° for the case
    without axial force). However it was noted that the lateral,
    central load position changed slightly for these beams. Thus,
    crack initiation was not due solely to mode II deformation.

5. The initial cracks from the notches, No. 1 in Fig. 35,
    stabilized when a central crack, No. 2, formed due to tensile
    stresses in the center of the beam exceeding the material
    strength. This crack eventually propagated through the beams
    with the final pattern shown in Fig. 35. A comparison with
    the CRACKER program is shown in Fig. 36. Increasing the axial
    force simply allowed for a higher vertical force until the
    net, central, tensile stress (calculated from the Brazilian
    formula) reached the material limit at which time crack No. 2
    formed.

Barr et al. (1987) developed and applied two compact test specimen
geometries suitable for studying the shear strength of concrete and
rocks. The first test geometry was a symmetrically notched compact
beam specimen loaded adjacent to the notches by forces which created a
concentrated shear zone in the near absence of bending moments. One
disadvantage of the geometry was that a large number of non-shear
fractures were obtained in experimental work. A numerical analysis of
the geometry showed that tensile zones are created in the areas
adjacent to the roots of the notches. These tensile stresses may be
the cause of some of the non-shear types of failure observed in
practice.

A numerical study was carried out on the various potential errors
in the use of the above geometry. Errors in the positioning of the
loads, errors in the notch depths and also errors in both the loads
and notch depths were investigated. The numerical results showed that
errors in the positioning of the applied loads were not significant.
On the other hand, errors in the notch depths can result in a bending
stress situation being created in the shear zone resulting in a non-
shear type of failure. In the case of the combined errors, the
greater influence was the error in the loading position for shallow
notches and errors in the notch depth for the deeper notches. The
main conclusion of the numerical study was that all three types of
errors investigated can result in a non-shear failure in the test
specimens. Thus great care is essential when using this test
geometry.

In view of the above results, a circumferentially notched compact
shear test specimen geometry was developed (Fig. 37). This test
specimen geometry gave reproducible results with a low coefficient of
variation - generally below 10%. The shear strengths reported for
various concrete mixes showed that the shear strength values were
independent of the notch depth used and also of the maximum size of
the coarse aggregate used in the mix. However, these results are
limited in scope and future work on size effects, using larger test
specimen sizes, is required before firm conclusions can be drawn
regarding the effects of the size of the coarse aggregates.

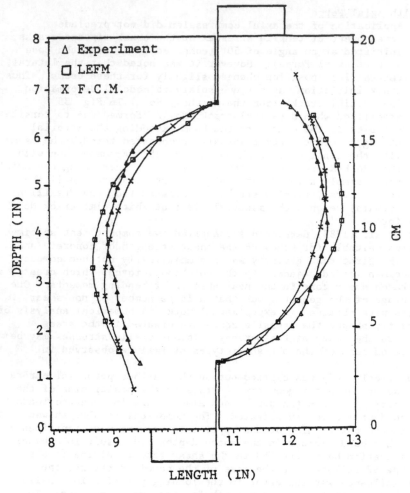

Fig. 34. Crack configuration for beam with no axial force.

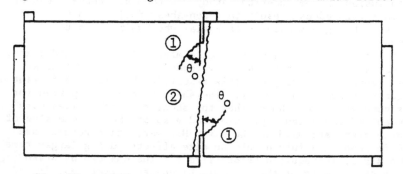

Fig. 35. Crack patterns for beam with axial force.

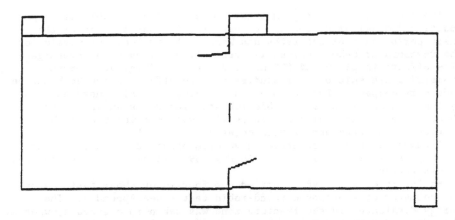

a. Cracks when adding the process zone
Fictitious Crack Model

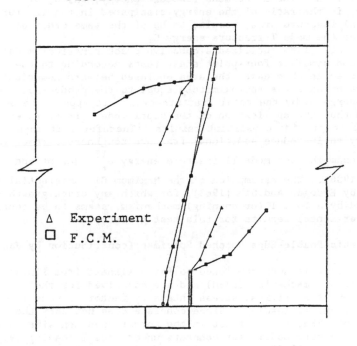

△ Experiment
□ F.C.M.

b. Crack configuration at failure

Fig. 36. Theoretical and experimental crack patterns for beam
with axial force.

The two test specimen geometries discussed are suitable for additional development and can be used for further studies of the shear performance of concretes and rocks. A more compact version of the rectangular beam specimen is currently being used to investigate size effects in plain and FRC mixes at Cardiff (Fig. 38). Both geometries are suitable for studies of size effects since both can be made more compact. Furthermore, the circumferentially notched cylindrical specimen is suitable for studying rocks subjected to shear. The cylindrical specimen is also suitable for use in the testing of concrete and/or rock cores.

A number of other geometries have been proposed recently by Barr and Hughes (1989) and are shown in Fig. 39. These are currently under investigation.

Ballatore et al. (1989) carried out four-point shear tests using the geometry of a proposed round-robin test - see Appendix. The energy dissipated in the fracture zone was taken into account, whereas the energy dissipated by punching at the supports was deliberately neglected by means of a reference bar (Fig. 4 - Appendix) and other suitable devices. The mixed-mode fracture energy, which, by definition, is the ratio of the energy dissipated in the fracture zone to the total fracture area, results to be of the same order of magnitude as the mode I fracture energy G_F.

Bocca et al. (1989) utilized a cohesive crack numerical model to interpret and simulate four-point shear tests according to the proposal. Even in this case, the area enclosed between numerical curve and deflection axis was approximately equal to the product of mode I fracture energy G_F by the total fracture area. It appears to be remarkable that the application of the usual mode I fracture energy G_F, is able to provide consistent results. Therefore, it appears unnecessary to introduce additional fracture toughness parameters, like, for example, the mode II fracture energy G_F^{II}, Bazant and Pfeiffer (1986). The assumption of the Maximum Circumferential Stress Criterion by Erdogan and Sih (1963), for which any crack growth step is produced by a mode I (or opening) mechanism, seems to be confirmed by the experimental results for this test geometry.

3.4 Quadratic Double-Edge-Notched Specimen (contribution by Jan G. M. van Mier)

Most of the test methods mentioned so far originate from linear elastic fracture mechanics (LEFM) and are conceived for the determination of a critical stress intensity factor in mode II or mixed-mode I and II loading. Since concrete does not obey the conditions of LEFM, at least not in the normal specimen size range, non-linear fracture models for concrete under mode I loading (e.g. Hillerborg et al. 1976) are currently extended for mixed-mode loading. In this case a different test set-up is used, in which it is tried to isolate a uniform process-zone (defined as any array of microcracks in front of a traction free macrocrack) and to subject it to combined tension and shear, e.g. Reinhardt et al. (1989); Hassanzadeh et al. (1987); Keuser (1988). It is believed that from such tests, constitutive properties for the process-zone may be derived. In fact the

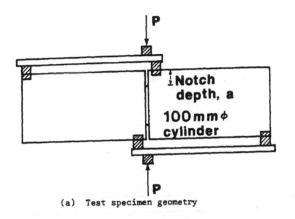

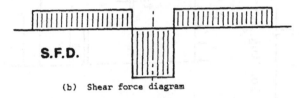

S.F.D.

(b) Shear force diagram

Fig. 37. Circumferentially-notched cylindrical shear test
specimens.

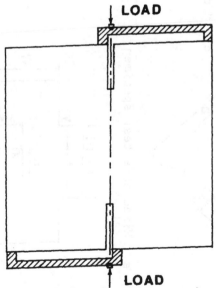

Fig. 38. Compact cube/prism shear test specimen used to study
size effects.

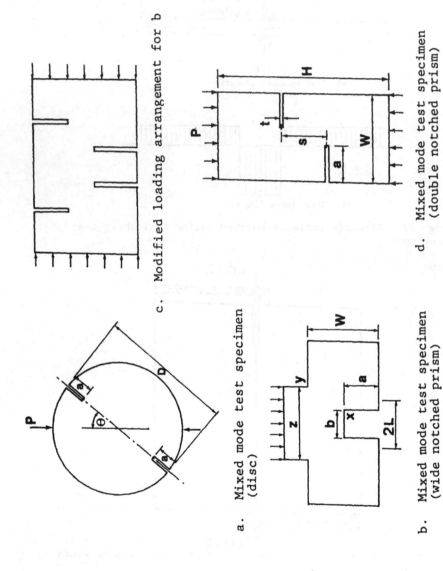

c. Modified loading arrangement for b

a. Mixed mode test specimen
(disc)

b. Mixed mode test specimen
(wide notched prism)

d. Mixed mode test specimen
(double notched prism)

Fig. 39. Other mixed-mode testing geometries.

experiments can be regarded as verification of aggregate interlock
theories for small crack widths.

This seems all rather simple and straightforward. Yet, after the
first experiments were carried out, it was found that the development
of a uniform process-zone under mode I loading is virtually impossible
Van Mier (1986). Also in uniaxial tension non-uniform cracking is
observed, and the shape of the stress-crack width relation depends on
the stiffness of the specimen outside the crack-zone and on the
boundary conditions in the experiment (e.g. fixed or rotating specimen
boundaries, see Van Mier (1990)). This implies that a direct
measurement of the stress-crack width relation for concrete is not
possible, and that it may be derived indirectly only. This is a
serious problem, as difficulties will increase as soon as a process-
zone is loaded under combined tension and shear.

As mentioned before, mixed-mode equipments for measuring the
constitutive properties of concrete were developed at different
laboratories, viz. in Lund, Hassanzadeh et al. (1987), Darmstadt,
Keuser (1988), and Delft, Reinhardt et al. (1989). The design and
mechanical functioning of the mixed-mode test rigs, as well as the
size of the specimens that can be tested in the respective machines
are completely different. At the Stevin Laboratory of Delft
University of Technology a biaxial test rig was developed in which a
separate control for mode I and mode II displacement in a mixed mode
test is possible, Reinhardt et al. (1989). In the current approach,
crack-growth under combined tension and shear is studied under well-
known boundary conditions. It is assumed that a crack initiates as
soon as the principal stress reaches the tensile strength of the
material.

The biaxial test rig consists basically of two independent rigid
frames that are fixed in an overall frame by means of plate springs.
An exploded view of the apparatus is shown in Fig. 40. One of the two
rigid frames consists of two coupled frames, and the second frame can
slide in between these coupled frames. The coupled frames can move in
the horizontal direction and are fixed to the overall frame in the
vertical direction. The middle frame can move vertically and is fixed
in the horizontal direction. To both the middle frame and the coupled
outer frames a load-cell and a hydraulic actuator, with a capacity of
100 kN in tension or compression, are connected as indicated. The
maximum load is however restricted to 50 kN. Square Double-Edge-
Notched (DEN) concrete plates of size 200 x 200 x 50 mm are loaded in
the apparatus. The experiments resemble the well known shear-box
experiments from rock mechanics.

The loading procedure is clarified in Fig. 41. In uniaxial
tension, a concrete plate is glued between the upper side of the
middle frame and the lower side of the coupled outer frames (Fig.
41a). When the glue has set, the specimen is loaded by moving the
middle frame upward. Since the coupled outer frames are fixed in the
vertical direction, a tensile stress develops in the concrete
specimen, which will eventually fracture the plate (Fig. 41b). The
concrete plates are double or single-edge-notched plates in order to
generate crack growth at a known location. This facilitates
deformation controlled testing. The same procedure is used for
applying a shear-load to a pre-cracked specimen, as shown in Fig. 42.

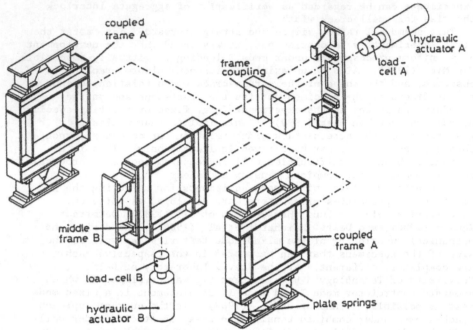

Fig. 40. Exploded view of the biaxial test rig of the Stevin Laboratory.

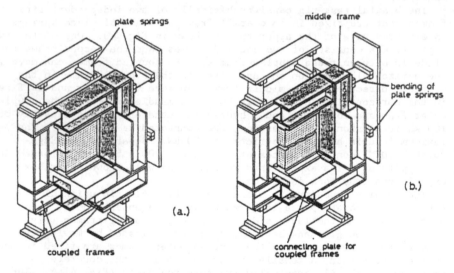

Fig. 41. Sectional view of the biaxial apparatus: at the beginning of an experiment (a), and after cracking (b).
Note the bending of the plate springs of the middle frame in Fig. b as a result of the upward movement of this frame with respect to the coupled frames.

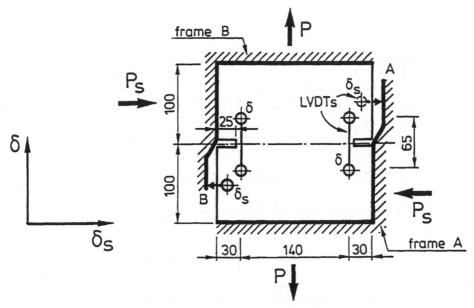

Fig. 42. Double-edge-notched square plate for mixed-mode tests.

Until now the cracking of concrete specimens was studied following different load-paths. Some of the results were reported by van Mier and Nooru-Mohamed (1988); van Mier (1989); Nooru-Mohamed and van Mier (1989); and van Mier et al. (1990), and here only examples of compressive shear tests on pre-cracked concrete specimens are shown. In these experiments, a specimen was subjected to lateral shear after pre-cracking of the plate to a prescribed axial crack width (respectively 50, 100 and 150 μm for the results presented in Fig. 43). The crack-pattern observed in the experiment with the average axial crack-opening of 150 μm is shown in Fig. 43c: crack ABC developed during pre-cracking (the axial load-displacement diagrams are shown in Fig. 43a.), crack CD during shearing (the shear load-displacement diagrams are shown in Fig. 43b). The boundary condition in the axial (vertical) direction during shearing was that the axial load was kept constant as a small compressive confinement of -1 kN. So far the results indicate that principal stress rotations are very important, especially during the application of the lateral shear load, and assumptions made in the so-called smeared rotating crack models are confirmed. For further test-results, the reader is referred to the previous publications.

3.5 Arcan Specimen
The device, shown in Fig. 44, was first proposed for use with fiber-reinforced, composite materials by Arcan et al. (1978). Extensive numerical and photoelastic studies were conducted on the entire system, Arcan et al. (1978); Arcan and Banks-Sills (1982); Banks-Sills and Bortman (1986), and it was demonstrated that a pure shear field (almost) surrounds the notch in the specimen shown in Fig. 44a. The

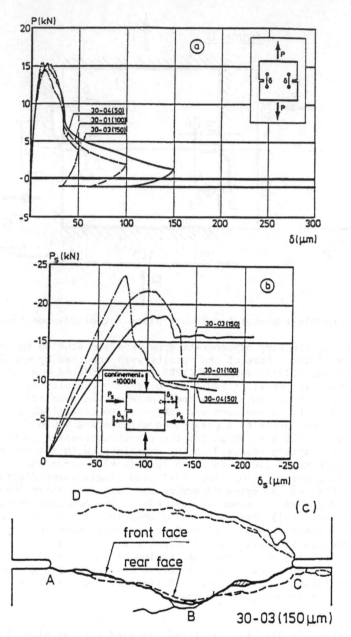

Fig. 43. P-δ (a) and P_s-$δ_s$ (b) diagrams for three compressive shear
tests with compressive confinement normal to the crack plane. The
tests were carried out at different values for the axial crack opening. The crack pattern for the experiment with an axial crack
opening of 150 μm is shown in Fig. c.

versatility of this arrangement is shown in Fig. 44b, where it is seen that mode I loading can be achieved. By rotating the holder, mixed-mode combinations of shear-tension or shear-compression may be obtained. This device is similar - but not identical - to one proposed by Isumi et al. (1986). The Arcan device has not been used to evaluate concrete or rock to our knowledge. Professor Arcan, in private correspondence, has suggested dimensions of a = 120 mm, b = 720 mm, t = 100 mm for the test section if the maximum aggregate size is about 20 mm. With these dimensions, the holding device becomes very large and heavy. However, it should be no more difficult to construct this device than the other devices or systems described in this report. This device deserves strong consideration as a tool to investigate mixed-mode fracture in concrete and rock in a systematic and controlled manner.

3.6 Semi-Circular Bend Specimen

This specimen geometry, shown in Fig. 45, was developed by Chong and his associates, Chong and Kuruppu (1984); Chong et al. (1989), for use with rock core samples. It consists of a disk cut from the core and then cut in half to form two bending test specimens. The central notch may be vertical, $\phi = 90°$, for mode I testing or inclined for mixed-mode testing. Another possible mixed-mode test geometry is with the notch off-center. This geometry has been analysed extensively and numerous tests conducted on rock samples. One major use for this would be to obtain fracture parameters for in-situ concrete from cores.

Because of the relatively small size, in the order of 100-150 mm diameter, correlations would need to be made with fracture properties obtained from other tests on the same material. The simplicity and ease of use make this an attractive specimen.

3.7 Round Robin Test Specimen

The mixed-mode fracture testing geometry conceived by Iosipescu (1967) and adapted for use with concrete by Arrea and Ingraffea (1981), Bazant and Pfeiffer (1985a, 1985b); Swartz et al. (1988b); Swartz and Taha (1989); Ballatore et al. (1989); and Bocca et al. (1989) forms the basis for a proposed experimental round robin test.

The round-robin testing program being conducted by subcommittee C seeks to obtain detailed test data using the geometry shown in the Appendix with some variations in dimensions. In order for these tests to provide the necessary information to evaluate the suitability of the many models presented in this report the following should be done.

1. All geometry details, including that of supporting fixtures should be reported.
2. The requirement that all tests be conducted on a servo-controlled machine using a displacement feedback (δ_2 has been proposed) should be strictly followed.
3. The crack trajectories should be monitored and related to the load-displacement traces.
4. Crack width measurements should be made, if possible. Full field data such as obtainable from Moiré interferometry is desirable.

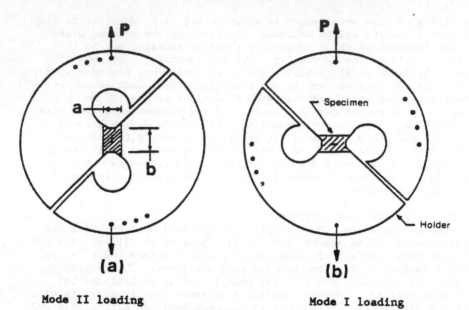

Mode II loading   **Mode I loading**

Fig. 44.   The Arcan specimen and holder.

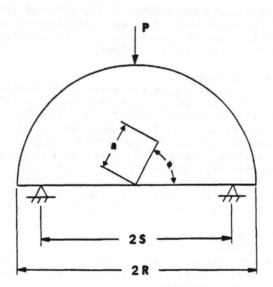

Fig. 45.   Semi-circular bend mixed-mode test specimen.

It is noted that such effects as aggregate-interlock or shear-friction will not be accounted for explicitly by these tests. Based on prior studies, it also seems clear that it will not be possible to verify the existence of a mode II fracture parameter such as fracture toughness.

4. Fracture of Concrete in Compression (contribution by Jan G.M. van Mier)

Fracture of concrete under compressive loadings is often regarded as the field of continuum mechanics. However, similarities with fracture mechanics can be drawn, especially when the localization of deformations in shear-bands is considered. In the past, fracture of concrete loaded in compression (crushing) was considered as a distributed phenomenon and the governing state variables were stresses and strains. Recently, however, it has become clear that localization of deformations occurs in uniaxial compression too, van Mier (1984), see Fig. 46. The main conclusion from this work is that the fracture of concrete in compression is a structural property, and that a shear band, which can be considered as a blunt crack with frictional resistance, traverses the specimen's cross-section when the load-deformation diagram describes a descending branch, van Mier (1986). The fracture process is greatly affected by boundary condition influences, viz. by frictional restraint at the specimen-loading platen interface, Kotsovos (1983); Vonk et al. (1989), see Fig. 47. Modeling of the localized fracture of concrete in compression can be done either by means of fracture mechanics, Li (1987), or by using the recently developed series coupling model for localization, Bazant (1989). In this latter model, a characteristic length of the material must be taken into account. It was shown that the characteristic length of the material is approximately equal to the characteristic length of the non-local continuum, which demonstrates that local continuum theories are not applicable to concrete.

In the above mentioned global approach, the fracture of the complete specimen is considered. When the microscopic aspects of compressive fracture of concrete are studied, cracking at the local level must be modelled. In this local approach, or micro-mechanics approach, the behavior of a specimen is explained by considering in detail the fracture processes at the grain-level. Since the 1960s it has become clear that extensive micro-cracking occurs in specimens subjected to compressive loading, Hsu et al. (1963). The weakest link is the aggregate-cement matrix bond, Taylor and Broms (1964), where tensile micro-cracks appear first. The effects of micro-cracking on the global specimen response have been studied in detail in the past, e.g., Budiansky and O'Connell (1976); Horii and Nemat-Nasser (1985); and Krajcinovic and Fonseka (1981). For compressive loading, the micro-cracks inclined to the compressive loading direction are often considered as the representative mode, Tasdemir et al. (1989); Scavia (1989)(see Fig. 48). The assumption is made that these microcracks nucleate at the interface of the aggregates and the cement matrix, and propagate into the matrix. In Tasdemir et al. (1989), a modified LEFM model was proposed which was used for determining the kink initiation stresses and crack initiation angles. The analytical results were in

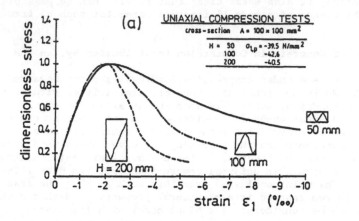

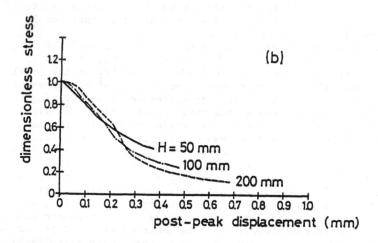

Fig. 46. Localization of fracture in uniaxial compression. In (a),
stress-strain curves for prisms with various height and constant
cross-sectional area are shown. In (b), the stress-post-peak
displacement curves derived from the same experiments. The total
post-peak displacement is independent of the length of a specimen.

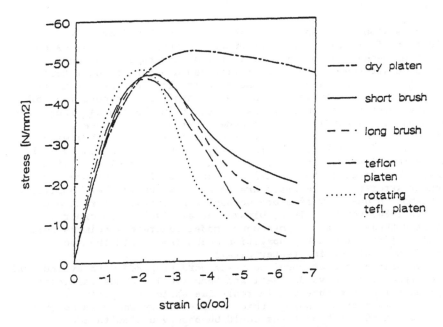

Fig. 47. Influence of the frictional restraint of different loading platens on the stress-strain curve of concrete in compression.

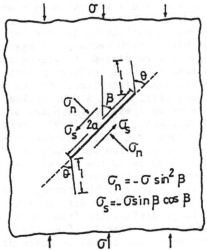

$$\sigma_n = -\sigma \sin^2 \beta$$
$$\sigma_s = -\sigma \sin \beta \cos \beta$$

Fig. 48. Inclined crack in a compressive stress field: initial crack, stresses at the onset of interface debonding, and kinks.

good agreement with experimental observations, Tasdemir et al. (1989); Maji, et al. (1989).

4.1 Compression Tests on Model Concrete (Contributed by A. K. Maji)

Application of fracture mechanics under compression is complicated by the various mechanisms of crack initiation and propagation under compressive stresses. Crack initiation could occur at the interface due to its inherent weakness. Cracks could also initiate from pre-existing defects, shrinkage cracks, etc. Arbitrary orientations of aggregate interface and mortar cracks result in the cracks being under mixed-mode stress condition where mode II or shear stresses are significant. These mode II shear stresses could exist along with tensile, mode I stresses. Alternatively, compressive stresses with negative mode I stress intensity factors could cause closure of the cracks. The effect of the normal stresses (tensile or compressive) and shear stresses on the crack are coupled, which further complicates any attempt to quantify fracture mechanics parameters (Bazant and Gambarova 1984; van Mier 1988; Divakar et al. 1987). In order to understand these different phenomena, model concrete specimens were studied. Different crack propagation mechanisms could then be isolated and analysed in greater detail.

Rectangular blocks of model concrete were prepared with cylindrical shaped aggregates and voids (Maji and Shah 1987). Crack initiation and propagation were observed in real-time while the specimens were loaded in compression. Holographic Interferometry was used so that the evolution of crack patterns could be observed simultaneously on the entire specimen with a high degree of sensitivity. In this technique, cracking and debonding are visible as abrupt breaks or kinks in the fringe patterns observed on the holograms. Progressive interface debonding and mortar crack growths were monitored at different stages of loading.

In order to obtain more quantitative information on the phenomena observed above, mortar blocks with rectangular aggregates were cast. The rectangular aggregates were placed at various inclinations to the applied compressive stress to obtain different ratios of normal and shear stresses on the aggregate interface. Cracks were observed to initiate at the interface (Fig. 49). A Mohr-Coulomb friction model was found to be capable of determining the onset of interface failure (Tasdemir et al. 1990; Tasdemir; et al. 1989). Mortar cracks initiated from the interface (Fig. 50). Stress intensity factors for progressive mortar crack propagation were calculated by a finite element analysis which incorporated the friction at the aggregate interface caused by compressive stresses. These stress intensity factors were found to vary widely if the propagating crack is assumed to be traction free. Normal and shear displacements measured across the crack also showed large differences from those calculated from FEM analysis of the specimen with a traction-free crack.

Similar observations were made on mortar crack propagation from an inclined notch. These specimens were similar to the specimens containing aggregates described above; except that the aggregates were replaced by inclined notches of various orientations (Maji et al. 1989). Typical holographic fringe patterns for a notch inclined at 72° are shown in Fig. 51. Holographic Interferometry was used to

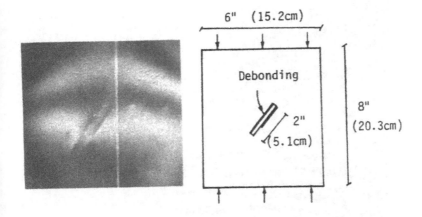

Fig. 49. Debonding of aggregate interface.

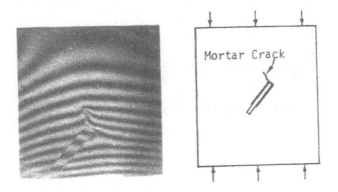

Fig. 50. Mortar crack initiation from interface.

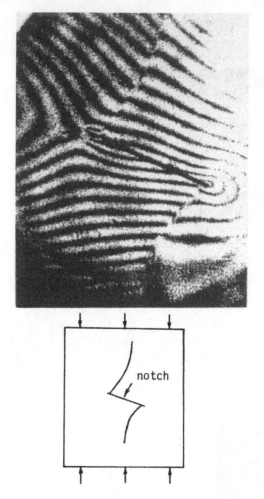

Fig. 51.   Crack propagation from notch.

measure crack opening and sliding along the entire propagating crack using a multiple observation method (Maji and Shah 1990). Although crack initiation was in mixed-mode, cracks curved in the direction of applied compressive load and the mode I stress intensity factor was dominant. Based on the FEM calculations, conclusions similar to the section above were obtained.

Traction forces on the crack faces were calculated such that the experimentally observed crack opening and sliding profiles matched those obtained from an FEM analysis incorporating those tractions. The mode I stress intensity factors corrected to include these traction forces were found to be within a reasonable band for both the aggregate specimens and the notch specimens mentioned above. It is expected that more detailed experimental and constitutive knowledge of crack faces subjected to mixed-mode stresses will yield better results. Therefore it is possible that the fracture mechanics approach could be capable of predicting crack initiation and propagation under compressive stresses leading to a fracture mechanics based model for compressive failure.

4.1 Analysis of Cracks Located in Compressive Stress Fields (contributed by Claudio Scavia)

Scavia (1990) has developed a numerical technique, based on the Displacement Discontinuity Method (DDM), for K_I and K_{II} evaluation under mixed-mode conditions. As to this problem, the DDM has been suitably modified by introducing at the crack tips a special element, with a parabolic variation of the displacements discontinuities, in order to reproduce the $r^{-1/2}$ stress singularity. The method has then been applied to the analysis of a series of fracture mechanics problems and a good degree of accuracy has been found, even using a very limited number of elements (e.g., using three elements an error less than 5% is made in the K_I and K_{II} computation). As the assumption of open crack, Scavia (1990) seldom applies when the crack is located in a compressive stress field, another type of displacement discontinuity element has been developed and introduced into the numerical technique, Scavia (1989). This element is such that:

(1) in the normal direction no displacement discontinuities exist (K_I=0);
(2) in the tangential direction, when a shear stress is applied, a shear resistance develops, according to an elastic-ideally plastic behavior of the crack surfaces.

In such a way, the reduction in the computed K_{II}, due to the development of frictional resistance, has been evaluated. Moreover, a propagation criterion for a crack subjected to normal and tangential stresses acting to infinity has been sought. Unlike for open cracks in mixed-mode conditions, the relationship between the applied critical stress intensity factors K_I^* and K_{II}^* has been found to be linear and given by:

$$K_{II}^* = K_{IIc} + K_I^* \tan\phi \qquad (61)$$

where:
$$K_{II}^* = \tau^* (\pi a)^{1/2}, \quad K_I^* = \sigma^*(\pi a)^{1/2}$$

$\tau*, \sigma*$ — tangential and normal compressive stresses, acting
at infinity, making the crack propagate;
K_{IIc} — critical value of the mode II Stress-Intensity Factor;
a — crack half length;
ϕ — frictional angle of the crack surface.

This result confirms the propagation criterion proposed by
Carpinteri (1986) on the basis of theoretical considerations.

5 Conclusions

This report has presented a survey of a recent, and growing, body of
work directed to determining the effects of mixed-mode loadings on
crack propagation in concrete. After reviewing a number of methods
for evaluating the state of stress in the region of a growing crack
and after presenting a large variety of proposed testing geometries
the committee feels a number of questions remain unresolved, that is
there is not yet a consensus of agreement on the following points.

1. Is the crack growth only by a mode I deformation mechanism?
    Would this be true at all size scales or might mode II
    deformation also be important at, say, the meso or micro
    scale?
2. What roles do aggregate interlock and shear friction play in
    the fracture process? Can these be treated by fracture
    mechanics methods?
3. Is the magnitude of mixed-mode fracture energy the same or
    different from mode I fracture energy?
4. What influence does the loading sequence have on crack
    propagation? Can this be modeled by methods of fracture
    mechanics?

6 References

Arcan, M., Hashin, Z. and Voloshin, A., (1978). A method to produce
    uniform plane-stress states with applications to fiber-reinforced
    materials, **Exp. Mech.**, 18, 4, 141-46.
Arcan, M. and Banks-Sills, L., (1982). Mode II fracture specimen-
    photoelastic analysis and results. In **Proceedings of the Seventh
    Inter. Conf. on Exp. Str. Anal.**, Haifa, Israel, Aug. 23-27, 187-
    201.
Arrea, M. and Ingraffea, A. R., (1981). Mixed-mode crack propagation
    in mortar and concrete. Report 81-13, Dept. of Structural
    Engineering, Cornell Univ., Ithaca, NY.
Ballatore, E., Carpinteri, A., Ferrara, G. and Melchiorri, G., (1990).
    Mixed mode fracture energy of concrete. **Engg. Fr. Mech.**, 35, 145-
    57.
Banks-Sills, L. and Bortman, Y., (1986). Mixed mode fracture
    specimens: analysis and testing. **Int. J. Fracture**, 30.
Barr, B. and Hughes, T. G., (1989). A comparative study of three
    proposed geometries used to investigate the fracture behavior of
    materials under mixed mode loading. In R3, 59-65.

Barr, B. and Hughes, T. G., (1988). Numerical study of shear (mode II) type test specimen geometries. In R2, 313-24.

Barr, B., Hasso, A. and Khalifa, S., (1987). A study of mode II (shear) fracture of notched beams, cylinders and cores. In SEM-RILEM International Conference on Fracture of Concrete and Rock, ed. S. P. Shah and S. E. Swartz, Houston, TX, June 17-19, 370-82.

Bazant, Z. P. and Gambarova, P. (1984). Crack shear in concrete: crack band microplane model. Journal of Structures Div., ASCE, 110.

Bazant, Z. P., (1984). Size effect in blunt fracture: concrete, rock, metal. Journal of Engineering Mechanics, ASCE, 110, 518-35.

Bazant, Z. P. and Pfeiffer, P. A., (1985a). Comment on Ingraffea and Panthaki's analysis of shear fracture tests of concrete. In Finite Element Analysis of Reinforced Concrete Structures, Tokyo, Japan, 174-83.

Bazant, Z. P. and Pfeiffer, P. A., (1985b). Test of shear fracture and strain softening in concrete. 2nd Symposium on the Interaction of Non-nuclear Munitions on Structures, Panama City Beach, FL, 254-64.

Bazant, Z. P. and Pfeiffer, P. A., (1986). Shear fracture tests of concrete, Materials and Structures (R.I.L.E.M.), 19, 111-21.

Bazant, Z.P., (1989). Identification of strain-softening constitutive relation from uniaxial tests by series coupling model for localization. Cement and Concrete Research, 19(6), in print.

Bocca, P., Carpinteri, A., and Valente, S., (1990). Size effects in the mixed mode crack propagation: softening and snap-back analysis. Engg. Fr. Mech. 35, 159-170.

Broek, D., (1986). Elementary Engineering Fracture Mechanics, Martinus Nijhoff, The Hague.

Budiansky, B. and O'Connell, R.J., (1976). Elastic moduli and a cracked solid. Int. J. Solids & Struct., 12, pp. 81-97.

Carpinteri, A., Di Tommaso, A. and Viola, E., (1979). Collinear stress effect on the crack branching phenomenon. Materials and Structures (R.I.L.E.M.), 12, 439-46.

Carpinteri, A., (1981). Static and energetic fracture parameters for rocks and concretes. Materials and Structures (R.I.L.E.M.), 14, 151-62.

Carpinteri, A., (1982). Notch sensivity in fracture testing of aggregative materials. Engineering Fracture Mechanics, 16, 467-81.

Carpinteri, A., (1985). Interpretation of the Griffith instability as a bifurcation of the global equilibrium. Application of Fracture Mechanics to Cementitious Composites. N.A.T.O.-A.R.W., Sept. 4-7, 1984, Northwestern University, edited by S. P. Shah, Martinus Nijhoff Publishers, 287-316.

Carpinteri, A., (1986). Mechanical damage and crack growth in concrete. Martinus-Nijhoff Publishers, Dordrecht.

Carpinteri, A., Di Tommaso, A. and Fanelli, M., (1986). Influence of material parameters and geometry on cohesive crack propagation. Fracture Toughness and Fracture Energy of Concrete, Oct. 1-3, 1985, Lausanne, edited by F. H. Wittmann, Elsevier, 117-35.

Carpinteri, A., (1987). Stress-singularity and generalized fracture toughness at the vertex of re-entrant corners. Engineering

Fracture Mechanics, 26, 143-55.

Carpinteri, A., Ferrara, G. and Melchiorri, G. (1989). Single edge notched specimen subjected to four point shear: an experimental investigation. In **Fracture of Concrete and Rock-Recent Developments**, S. P. Shah, S. E. Swartz and B. Barr, eds., Elsevier Science Publishers, London/New York, pp. 605-14.

Chong, K. P. and Kuruppu, M. D., (1984). New specimen for fracture toughness determination of rock and other materials. Int. J. **Fracture**, 26, R52-R62.

Chong, K. P., Kuruppu, M. O., and Kuszmaul, J. S., (1989). Fracture toughness determination of rocks with core-based specimens. In R1, 13-25.

Di Leonardo, G., (1979). Fracture toughness characterization of materials under multiaxial loading. Int. **Journal of Fracture**, 15, 537-52.

Divakar, M. P., Fafitis, A. and Shah, S. P. (1987). A constitutive model for shear transfer in cracked concrete. **Journal of the Structural Div.**, ASCE, 113, 5, May.

Dutron, P. (1974). **Matériaux et Constructions**, 7.

Erdogan, F. and Sih, G. C., (1963). On the crack extension in plates under plane loading and transverse shear. J. **Basic Engg.**, 85, 519-27.

Hassanzadeh, M., Hillerborg, A., Zhou, F. P., (1987). Tests of material properties in mixed mode I and II. In **SEM-RILEM International Conference on Fracture of Concrete and Rock**, ed. S. P. Shah and S. E. Swartz, Houston, TX, June 17-19, 353-58.

Hillerborg, A., Peterson, P.E., Modeer, M., (1976). Analysis of crack formation and crack growth in concrete by means of fracture mechanics and finite elements. **Cement and Concrete Research**, 6, pp. 773-82.

Horii, H. and Nemat-Nasser, S., (1985). Compression-induced microcracked growth in brittle solids: axial splitting and shear failure, J. **Geophys. Res.**, 90, pp. 3105-25.

Hsu, T.T.C., Slate, F. O., Sturman, G. M. and Winter, G. (1963). Microcracking of plain concrete and the shape of the stress-strain curve. **ACI Journal**, 60(2), pp. 209-24.

Ingraffea, A. R. and Panthaki, M. J., (1985). Analysis of shear fracture tests of concrete beams. In **Finite Element Analysis of Reinforced Concrete Structures**, Tokyo, Japan, 151-73.

Ingraffea, A. R., (1985). Non-linear fracture models for discrete crack propagation. Application of Fracture Mechanics to Cementitious Composites, N.A.T.O.-A.R.W., Sept. 4-7, 1984, Northwestern University, edited by S.P. Shah, Martinus Nijhoff Publishers, 247-85.

Iosipescu, N., (1967). New accurate procedure for single shear testing of metals. J. **Mater.**, 2, 537-66.

Izumi, M., Mihashi, H. and Nomura, N., (1986). Fracture toughness of concrete for mode II. In **Fracture Toughness and Fracture Energy of Concrete**, ed F. H. Wittmann, Elsevier, Amsterdam, 347-54.

Jenq, Y. S. and Shah, S. P., (1985). Two Parameter Fracture Model for Concrete. J. **Engg. Mech.**, 111, 10, 1227-41.

Jenq, Y. S. and Shah, S. P., (1988). Mixed-mode fracture of concrete. Int. J. **Fracture**, 38, 123-42.

Keuser, W., (1988). Mixed-mode testing of plain concrete. **Darmstadt Concrete**, Vol. 3, pp. 57-66.

Kokini, K., Marangoni, R., Dorogy, G., and Ezzat, H., (1987). Crack propagation under mode II loading: an effective stress intensity factor method. **Engg. Fr. Mech.**, 28, 1.

Kotsovos, M.D., (1983). Effect of testing techniques on the post-ultimate behavior of concrete in compression, **Materials & Structures**, RILEM, Vol. 16, No. 91, pp. 3-12.

Krajcinovic, P. and Fonseka, G.K., (1981). The continuous damage theory of brittle materials, part I: general theory. J. **Appl. Mech.**, Trans. ASME, 48, pp. 809-15.

Kumosa, M. and Hull, D., (1987). Mixed mode fracture of composites using Iosipescu shear test. **Int. J. Fracture**, 102.

Li, V.C., (1987). Mechanics of shear rupture applied to earthquake zones. In **Fracture Mechanics of Rock**, Academic Press Inc. (London), pp. 351-428.

Maiti, S. K. and Smith, R. A., (1984). Comparison of the criteria for mixed mode brittle fracture based on the preinstability stress-strain field, part II, pure shear and uniaxial compressive loading. **Int. J. Fracture**, 24.

Maji, A. K. and Shah, S. P. (1987). Application of acoustic emission and laser holography to study microfracture of concrete. In **Nondestructive Testing of Concrete**, H. S. Lew, ed., ACI-SP112.

Maji, A.K., Shah, S.P. and Tasdemir, M.A., (1989). A study of mixed-mode crack propagation in mortar using holographic interferometry. In R3, pp. 210-17.

Maji, A. K. and Shah, S. P. (1990). Measurement of crack profiles by holographic interferometry. **Exp. Mech.**, 30, 2, 201-7.

Melin, S. (1986). When does a crack grow under mode II conditions? **Int. J. Fracture**, 30.

Nooru-Mohamed, M.B. and Van Mier, J.G.M., (1989). Fracture of concrete under mixed-mode loading. In **Fracture of Concrete and Rock - Recent Developments**, S.P. Shah, S.E. Swartz and B. Barr, ed., Elsevier Science Publishers, London/New York, pp. 458-467.

Nuismer, R. J., (1975). An energy release rate criterion for mixed mode fracture. **Int. J. Fracture**, 11, 245-50.

Peterson, P. E., (1981). Crack growth and development of fracture zones in plain concrete and similar materials. Report TVBM-1006, Division of Building Materials, Lund Institute of Technology, Sweden.

Reinhardt, H. W., Cornelissen, H. A. W. and Hordijk, D. A., (1989). Mixed mode fracture tests on concrete. In R1, 117-30.

Rots, J. G. and de Borst, R., (1987). Analysis of mixed mode fracture in concrete. J. **Engg. Mech.**, 113, 11, 1739-58.

Scavia, C., (1989). Analysis of crack propagation in a compressive stress field. In **Fracture of Concrete and Rock - Recent Developments**, S.P. Shah, S.E. Swartz and B. Barr, eds., Elsevier Science Publishers, London/New York, pp. 635-44.

Scavia, C., (1990). Fracture mechanics approach to stability analysis of rock slopes. Accepted for publication in **Engineering Fracture Mechanics**.

Sih, G. C., (1973). Some basic problems in fracture mechanics and new concepts. **Engineering Fracture Mechanics**, 5, 365-77.

189

Sih, G. C. and Macdonald, B., (1974). Fracture mechanics appied to engineering problems - strain energy density fracture criterion. **Engineering Fracture Mechanics**, 6, pp. 361-86.

Swartz, S. E., Lu, L. W. and Tang, L. D., (1988a). Mixed-mode fracture toughness testing of concrete beams in three-point bending. **Materials and Structures**, 21, 33-40.

Swartz, S. E., Lu, L. W., Tang, L. D. and Refai, T. M., (1988b). Mode II fracture parameter estimates for concrete from beam specimens. **Exper. Mech.**, 28, 2, 146-53.

Swartz, S. E. and Taha, N. M., (1990). Mixed mode crack propagation and fracture in concrete. **Engg. Fr. Mech.**, 35, 137-44.

Swenson, D. V., (1986). Modeling mixed-mode dynamic crack propagation using finite elements: theory and applications. Ph.D. dissertation, Cornell Univ., Ithaca, NY.

Taha, N., (1988). Mixed mode fracture of four-point loaded beams. Ph.D. Thesis, Civil Engg. Dept., Kansas State University, Manhattan, KS.

Tasdemir, M.A., Maji, A.K., and Shah, S.P., (1989). Mixed-mode crack propagation in concrete under uniaxial compressive loading. In **Fracture of Concrete and Rock - Recent Developments**, S.P. Shah, S.E. Swartz and B. Barr, eds., Elsevier Science Publishers, London/New York, pp. 615-24.

Tasdemir, M. A., Maji, A. K. and Shah, S. P. (1990). Crack initiation and propagation in concrete under compression. Journal of **Engineering Mechanics**, ASCE, May.

Taylor, M.A., and Broms, B.B., (1964). Shear bond strength between coarse aggregate and cement paste or mortar. **ACI Journal**, 61, pp. 937-57.

Van Mier, J.G.M., (1984). Strain-softening of concrete under multiaxial loading conditions, Ph.D. Dissertation, Eindhoven University of technology, Eindhoven, the Netherlands, November, 349 pp.

Van Mier, J.G.M., (1986). Fracture of concrete under complex stress. **HERON**, 31(3), 90 pp.

Van Mier, J. G.M. (1988). Fracture study of concrete specimens subjected to combined tensile and shear loading. Int. Conf. on **Measurement and Testing in Civil Engineering**, Lyon, Sept.

Van Mier, J. G. M. and Nooru-Mohamed, M. B., (1988). Fracture of concrete under tensile and shear-like loadings. In R2, 433-47.

Van Mier, J. G. M., (1989). Mixed-mode fracture of concrete under different boundary conditions. In R3, 51-8.

Van Mier, J.G.M., (1990). Mode I behavior of concrete: influence of the rotational stiffness outside the crack-zone. In **Analysis of Concrete Structures by Fracture Mechanics, Proceedings** of the RILEM workshop dedicated to Prof. A. Hillerborg, Abisko, Sweden, June 1989; L. Elfgren ed., Chapman & Hall, in print.

Van Mier, J.G.M., Nooru-Mohamed, M.B., Schlangen, E., (1990). Experimental analysis of mixed-mode I and II behavior of concrete. In **Analysis of Concrete Structures by Fracture Mechanics, Proceedings** of the RILEM workshop dedicated to Prof. A. Hillerborg, Abisko, Sweden, June 1989, L. Elfgren ed., Chapman & Hall, in print.

Vonk, R., Rutten, H.S., Van Mier, J.G.M., and Fijneman, H.J., (1989). Influence of boundary conditions on softening of concrete loaded in compression. In **Fracture of Concrete and Rock - Recent Developments**, S.P. Shah, S.E. Swartz and B. Barr, eds., Elsevier Science Publishers, London/New York, pp. 711-20.

Wecharatana, M. and Shah, S. P., (1983). Prediction of nonlinear fracture process zone in concrete. **Journal of Engineering Mechanics ASCE**, 109, 123-46.

Zhou, Q., Gong, S. and Chen, Z., (1987). Compression shear fracture mechanism for marble. In **SEM-RILEM International Conference on Fracture of Concrete and Rock**, ed. S. P. Shah and S. E. Swartz, Houston, TX, June 17-19, 580-86.

R1. **Fracture of Concrete and Rock**, ed. S. P. Shah and S. E. Swartz, Springer-Verlag, New York Inc., 1989.

R2. **Preprints of Proceedings** of International Workshop on Fracture Toughness and Fracture Energy-Test Methods for Concrete and Rock, ed. H. Mihashi, H. Takahashi, and F. H. Wittmann, Sendai, Japan, October 12-14, 1988.

R3. **Proceedings** 1989 SEM Spring Conference on Experimental Mechanics, Cambridge, MA, May 28-June 1.

APPENDIX, Chapter 3

R.I.L.E.M. Committee 89-FMT
Fracture Mechanics of Concrete: Test Method
Subcommittee C: Mixed-Mode Crack Propagation

Round Robin

On the basis of the preliminary experimental investigation, Carpinteri et al. (1989), a round robin among different laboratories should be realized with two concretes, as will be shown in the following sections.

Materials

Cement. High strength Portland Cement (according to B.S.: R.H. Portland Cement; according to ASTM: type III) with strength on standard mortar of at least 50 Mpa (after 28 days).
Aggregates. A granulometric distribution is proposed for the two concretes with alluvial aggregates, substantially in agreement with the RILEM Preliminary Recommendation for the Composition and the Placing of a Standard Concrete, Dutron (1974), Fig. 1 and 2.
Additive. Do not use any additive.
Mix proportions. Water/cement = 0.48.

Size and Number of Specimens

Make reference to the testing geometry of Fig. 3 (four point shear specimen with a single initial crack). Four identical specimens should be tested for each case of Table 1. The mechanical behavior of concrete should be characterized by:
- compression strength (cubic or cylindric) after 28 days;
- static and dynamic elastic modulus after 28 days;
- mode I fracture energy G_F by T.P.B.T. according to the R.I.L.E.M. recommendation;
- mode I fracture parameters (i.e., critical stress intensity factor, K_{IC}^s, and critical crack tip opening displacement, $CTOD_c$) based on the proposed RILEM recommendation using three-point bend test, Jenq and Shah (1985, 1988).

Cure of Specimens

The specimens should be kept in a controlled environment at 20°C and 95% relative humidity up to testing. The notch should be performed by means of a circular saw (thickness $\leq$ 4 mm) after the 14th day.

Specimen Preparation

Each specimen should be provided with four steel supports of sizes 2 x 2 x 10 cm glued at the four intended loading points, and with displacement transducers (LVDT) to measure the following quantities (Fig. 4):

- two transducers to measure the deflections δ_1 and δ_2 of the two upper loading points;
- two transducers (one per each side of the beam, in position 3) to measure the CMSD.
- one transducer (in the center, position 4) to measure the CMOD.

Testing Procedure

The tests should be carried out using a servocontrolled machine and the deflection δ_2 as feed-back signal. The imposed deflection rate should be ~ 0.025 μm/sec.

The following diagrams should be recorded:

(1) $P - \delta_1$ and $P - \delta_2$, P being the total load, from which it is possible to obtain the curves $F_1 - \delta_1$ and $F_2 - \delta_2$, for the evaluation of the Mixed-Mode fracture energy.

(2) P - CMSD, where CMSD is the average between the two transducers on the two beam sides, in order to take into account torsional effects; P - CMOD.

(3) $P - \delta$ and P - CMOD for Mode I tests.

(4)   Crack trajectories.

Table 1.   Specimen dimension

D_{max} (mm)	Thickness t (cm)	Depth b (cm)	Length (cm)	Span 1 (cm)	Crack Length (cm)	c (cm)
10	10	10	44	40	2	4
	10	20	84	80	4	8
20	10	20	84	80	4	8

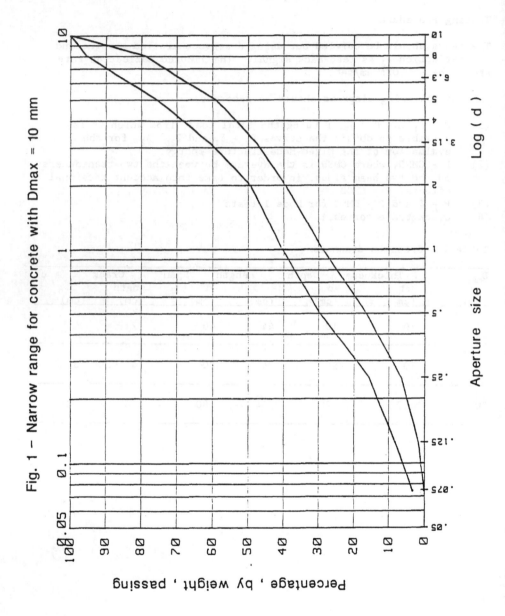

Fig. 1 – Narrow range for concrete with Dmax = 10 mm

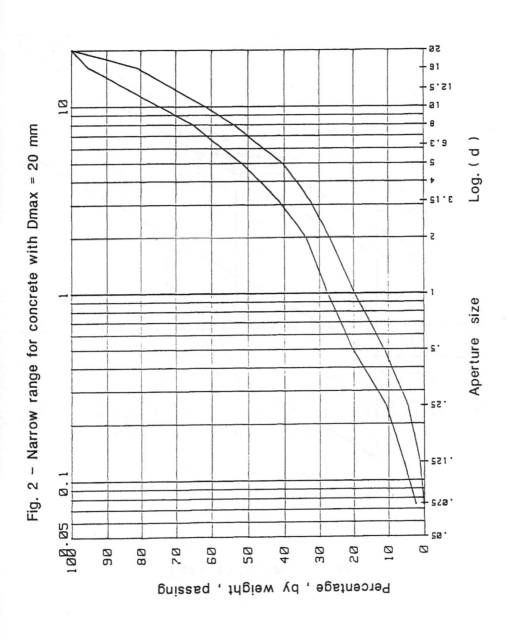

Fig. 2 – Narrow range for concrete with Dmax = 20 mm

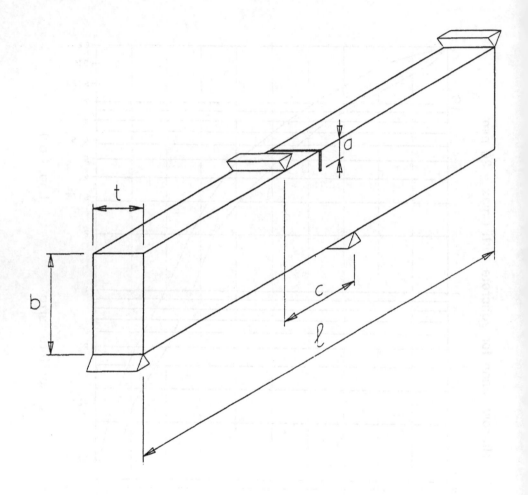

Fig. 3 – Testing geometry

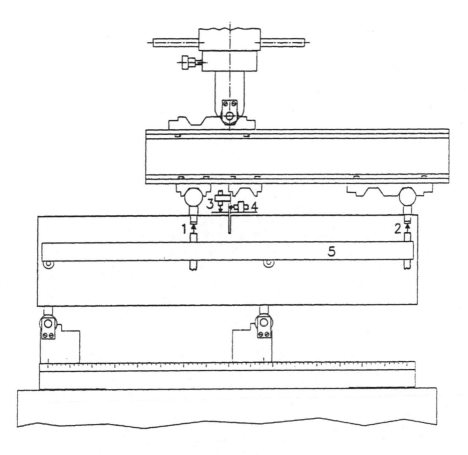

1 — LVDT for the measurement of δ_1 deflection

2 — LVDT for the measurement of δ_2 deflection

3 — LVDT for the measurement of C.M.S.D.

4 — LVDT for the measurement of C.M.O.D.

5 — Reference bar for δ_1 and δ_2 measurement

Fig. 4 — Loading configuration

4 LOADING RATE, TEMPERATURE, AND HUMIDITY EFFECTS

H.W. REINHARDT
University of Stuttgart, Stuttgart, Germany

1 Loading rate effects

1.1 Introduction
Experimental data have shown that a rate effect exists
which can increase strength and deformation of concrete. A
variety of testing methods has been used to show the rate
effect. However, one should be cautious interpreting
experimental results since other effects may also have a
significant influence. Inertia effects, local damage, or
stress wave reflection can be mentioned which should
receive due attention. The testing methods are differently
prone to such effects which requires a careful evaluation.
There are several questions concerning loading rate
effects with respect to testing. First, the testing method
must be clear and physically sound so that the quantity
which has to be determined can be measured. Second,
testing methods should be described or developed which are
suited to investigate the reason for the rate effect found.
An example may clarify the two aspects: in a bending test,
inertia effects must be considered in order to calculate
the true fracture energy, which is the first aspect.
Whether the increase of fracture energy at high loading
rates is caused by an increase of the process zone size or
by the forced fracture of aggregate or other causes should
be investigated by other techniques, which is the second
aspect.
In the following chapters, several loading configura—
tions will be examined with respect to possible errors in
evaluating testing results.

1.2 Bending test
The bending test is frequently used at high rates of
loading since the forces to fracture are moderate and can
be supplied rather easily. In metal testing, the Charpy
impact test on notched specimens, for instance, is widely
recommended for the determination of total fracture energy
and it is especially used for the detection of temperature
influences on the ductility of metals. There are, however,
limitations in the validity of the results since stiffness
of the pendulum, size and geometry of the specimen and
stiffness of the supports influence the results. Thus, in
metal testing the whole setup is standardized and the
remaining error may be acceptable.

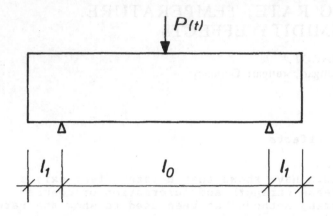

Fig. 1 Beam specimen

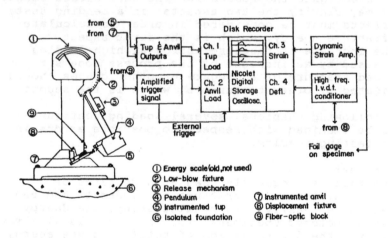

from ⑤
from ⑦

Disk Recorder

Tup & Anvil
Outputs

Ch. 1
Tup
Load

Ch. 3
Strain

Dynamic
Strain Amp.

from ⑨

Amplified
trigger
signal

Ch. 2
Anvil
Load

Nicolet
Digital
Storage
Oscillosc.

Ch. 4
Defl.

High freq.
l.v.d.t.
conditioner

External
trigger

from ⑧

Foil gage
on specimen

① Energy scale(old,not used)
② Low-blow fixture
③ Release mechanism
④ Pendulum
⑤ Instrumented tup
⑥ Isolated foundation
⑦ Instrumented anvil
⑧ Displacement fixture
⑨ Fiber-optic block

Fig. 2 Schematic of modified instrumented Charpy test

1.2.1 Inertia effect

Let us assume a beam of span l_o and with two parts of l_1 overhanging the supports is hit by a central force $P_t(t)$, Fig. 1. This beam will deform according to material properties and geometry. If there is an edge crack (notched beam) the deformation is likely to be linear with the maximum in the centre. If there is no crack or if the beam is reinforced the deflection line will be a higher order polynomial which may be approximated by a sine function. The inertia effects of both cases have been analysed by Bentur, Mindess and Banthia (1986). The inertia resistance detected by a force $P_1(t)$ is in case of constant cross-section and mass distribution:

$$P_1(t) = \rho A \ddot{u}_o(t) \; [\frac{l_o}{3} + \frac{8}{3} \frac{l_1^3}{l_o^2}] \tag{1}$$

for linear displacement distribution and

$$P_1(t) = \rho A \ddot{u}_o(t) \; [\frac{l_o}{2} + \frac{2\pi^2}{3} \frac{l_1^3}{l_o^2}] \tag{2}$$

for sinusoidal displacement distribution. A is cross-sectional area, ρ mass density of the material, u_o displacement at the centre of the beam, $\ddot{u}_o$ acceleration at the centre of the beam.

If the total load $P_t(t)$ and the acceleration $\ddot{u}_o$ are measured, the effective load which causes bending, is

$$P_b(t) = P_t(t) - P_1(t) \tag{3}$$

The time deflection curve $\delta(t)$ can be measured separately or can be obtained by integration of the acceleration signal. $P(t)$ and $\delta(t)$ lead to the fracture energy G_F. Suaris and Shah (1981) and Gopalaratnam, Shah and John (1983) have analysed the drop-weight-type and the Charpy-bending-type impact test analytically using a two-degree-of-freedom model (dof). The test beam is represented by a lumped mass with a certain stiffness and the hammer-tup assembly is represented by another lumped mass. The tup-specimen contact zone has an appropriate effective stiffness. The result of the analysis shows an increasing influence of inertia oscillations on the measured result with decreasing stiffness of the specimen and increasing stiffness of the contact area. The analytical peak load is given by

$$P_{t\,max} = v_o \; \sqrt{(k_e m_t)} \tag{4}$$

with v_o the striker (tup) velocity, k_e the contact stiffness, and m_t the lumped mass of the hammer-tup combination. The tup force oscillation is calculated by

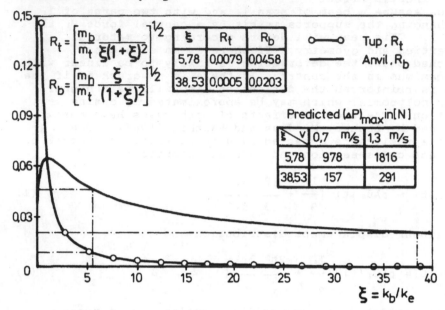

Fig. 3   Inertia influence on impact bending response

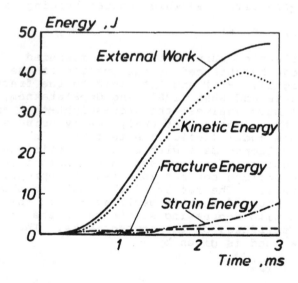

Fig. 4   Partition of energy in an impact test

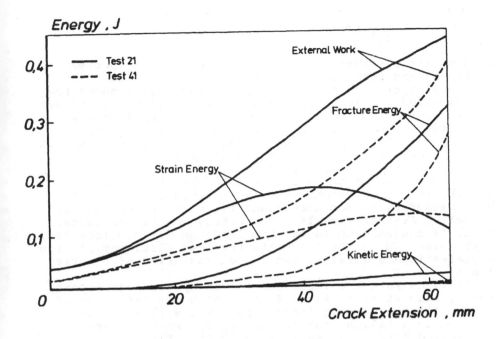

Fig. 5 Partition of energy in quasi—static testing

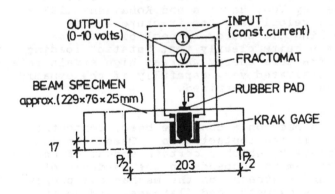

Fig. 6 KRAK gauge

$$R_t = \sqrt{\left(\frac{m_b}{m_t} \; \frac{1}{\zeta(1+\zeta)^2} \right)} \tag{5}$$

with m_b the specimen mass, ζ the ratio between specimen stiffness and contact stiffness. Fig. 2 shows the arrangement of the testing equipment and Fig. 3 shows how the amplitude ratio varies with the ratio of the stiffness. It can be seen how sensitive the recorded tup load reacts on a stiff beam—tup contact behaviour. The situation improves drastically if, for instance, a rubber or aluminium pad is placed between tup and specimen; however, the consequence would be a reduction in strain rate.

The analysis of the above is also appropriate for bending by a drop weight (Suaris and Shah 1983). It has been used for plain concrete and fibre reinforced concrete as well. From the results, the energy can be calculated which is consumed up to maximum load and failure load and, maybe also, up to the load when the load displacement line deviates from a straight line, i.e. some kind of elastic limit.

To illustrate the importance of an appropriate account for kinetic effects, Figs. 4 and 5 show the partition of energy for two bending tests, the first with an impact loading by a falling hammer, the second with a displacement controlled loading at smaller velocity. Fig. 4 from Kobayashi, Hawkins and Du (1989) shows the results of a finite element analysis of the test by Bentur, Mindess and Banthia (1986). It can be seen that the kinetic energy increases dramatically during the impact loading while fracture energy and strain energy are at a comparably low level.

Opposite to that, Fig. 5 shows the result of a test with a maximum strain rate of 0.1 s^{-1} (Test 21) and 0.005 s^{-1} (Test 41) performed by Yon, Hawkins and Kobayashi (1989a). At these rather low velocities, the fracture energy dominates and the kinectic energy is negligible. The examples demonstrate quite clearly that "static" loading leads directly to fracture energy whereas high strain rate loading has to be evaluated very carefully if the dynamic fracture energy has to be determined.

1.2.2 Local damage

During impact of a hammer on a concrete beam, concentrated stresses will occur in the contact area. These can lead to damage of the concrete by which a certain amount of energy is dissipated. This amount depends on the hardness of the hammer, its shape and flatness, on the mechanical properties of the beam, its geometry and flatness, and on the measure of coincidence of the hammer and beam surface. In order to avoid arbitrary influences on the results, the contact surfaces should fit as good and reproducible as possible and/or the contact surface of the beam should be covered by a pad always of the same material.

The contact force can be approximated by the Hertz formula

$$F = ka^{3/2} \tag{6}$$

where a is the penetration of the striker into the material and k is a "constant" which is a measure of the stiffness of the impact zone. k depends on the mechanical properties of the specimen and of the striker, on the loading intensity, and on the geometries involved. F and a depend on time. Values for k have been found in experiments ranging between 4 kN/mm$^{3/2}$ for a rubber pad, 24 kN/mm$^{3/2}$ for a 12 mm ply pad, and 200 kN/mm$^{3/2}$ for a steel plate (Hughes and Beeby 1982). But it is also true that k is not really a constant but varies with stress intensity. Tests at Delft University of Technology aim to establish k—values for various geometries and materials (Van Mier et al. 1990). Tests between colliding concrete bodies have shown k—values of 50 kN/mm$^{3/2}$ for a spherical/planar surface combination, about 1 for conical/planar, and 80 for truncated conical/planar combinations. The concrete quality had a rather low influence on these values.

Since k is not a constant, the analysis as referred to in the preceding Chapter can only be an approximation. Further studies are encouraged to improve the knowledge with respect to local phenomena in impact testing.

1.2.3 Instrumentation

Electrical and optical methods are appropriate to measure relevant quantities during impact testing. As concrete is concerned, optical methods can be used for the triggering of recording instruments or for crack occurrence measurements. A light beam. which is crossed by the drop weight of an impact testing equipment, can trigger a storage oscilloscope or disc recorder. A bundle of fibres cast into the specimen can detect cracking in the interior of a specimen (Rossi and Le Maour 1986, Rossi 1986). Cameras with high framing rate (> 10,000 frames per second) may be used for measuring displacement and crack propagation. However, since concrete is a softening material exhibiting a process zone, it is difficult to locate a crack tip properly; but on the other hand, a full—field picture always gives very useful information (Mindess et al. 1986).

Electrical devices use strain gauges, LDVTs, cracking gauges, load cells, accelerometers and load sensitive pads. Strain gauges follow all frequencies which occur at impact testing of concrete. LDVTs should be mounted very tightly to the surface in order to prevent all spureous oscillations (Reinhardt, Körmeling and Zielinski 1986). Cracking gauges are used successfully (John and Shah 1986) on bending specimens, see Fig. 6. The so—called KRAK gauges (TTI Division, Hartrum Corporation, Chaska, MN, USA)

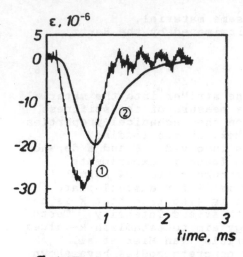

Fig. 7 Strain—time plots
1) Frequency band 0 to 500 kHz
2) Frequency band 0 to   1 kHz

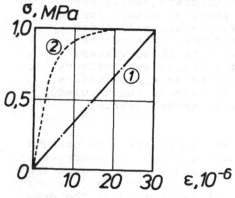

Fig. 8 Stress—strain plot
1) Stress and strain measured
   with 0 to 500 kHz
2) Strain measured with 0 to
   1 kHz

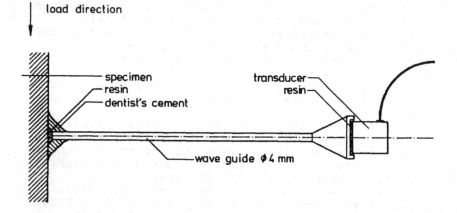

Fig. 9 Wave guide to avoid damage of transducer due to
        impact loading

are mounted with epoxy resin to the surface of the
specimen. Glue and backing material should be brittle in
order to follow a crack and not to bridge it.
Accelerometers must be mounted in a clearly defined
position which should exclude misalignment. Load cells
should be designed in such a way that the resonance
frequency is higher than the highest expectable frequency
of a test. An interesting alternative for load cells may
be polymer pressure gauges. They consist of a material
(polyvinylidene fluoride) which becomes piezoelectrically
and pyroelectrically active when subjected to a large
electric field (2 MV/cm at room temperature). Two about 12
μm thin foils of sensitive material are sandwiched between
two protective layers of polycarbonate (Bur and Roth 1985).
These gauges are mainly used for air pressure measurements,
but have also been tested in soils (Chung, Bur and Holder
1985).

The recording system is an essential part of the
measuring chain. It is essential that the relevant
measuring frequency is high enough to follow the event.
Fig. 7 shows two examples of a strain—time recording the
one received by a 1 kHz band width oscilloscope, the other
with a 500 kHz oscilloscope (Frank 1984). It can be seen
that peak strain and time of occurrence of peak strain are
badly determined by too low a frequency band. When stress
and strain are plotted together in a stress—strain plot,
Fig. 8, it appears that the erroneous strain measurement
leads to a large overprediction of Young's modulus and to a
curved shape of the line whereas the correct measurement
leads to a straight relation between stress and strain
which had to be expected. The concrete has actually been
loaded to a fraction of its compressive strength.

As far as concrete testing is concerned, there should be
a measuring frequency of at least 10 kHz. Oscilloscopes
with a carrying frequency of only 5 kHz are not useful.
Various publications have been found where storage
oscilloscopes with sampling frequencies larger than 1 MHz
have been applied.

1.2.4 Various test methods
Acoustic emission analysis is one possible method which has
been used by Curbach, Maliszkiewicsz and Eibl (1989b). The
application of this well—known method (see e.g. Diederichs,
Schneider and Terrien 1983) to high strain rates requires
some special arrangements regarding the registration and
storage of the acoustic signal.

Due to impact loading the surface of the specimen
receives high accelerations in the magnitude up to 2000 g
which makes it necessary to protect the transducer.
Therefore a so—called wave—guide, which carries acoustic
waves in a longitudinal direction to the transducer but
which is weak in lateral direction, has to be installed.
To avoid losses due to wave reflections this wave guide has
to be made of a lightweight material (aluminium) with an

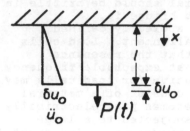

Fig. 10
Tensile specimen with displacement and acceleration distribution

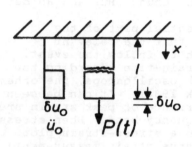

Fig. 11
Tensile specimen with discrete
cracking zone

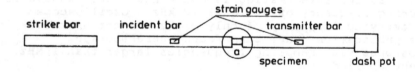

Fig. 12   Split Hopkinson bar

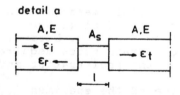

acoustic impedance of about 17 MNs/m³. The acoustic
impedance of concrete is about 9 MNs/m³. The principal
arrangement is shown in Fig. 9.

The analysis of the acoustic signal using traditional
equipment is not possible because of the short experiment
duration. Instead of this the original acoustic emission
signal has to be stored either in an analogue way on a
magnetic tape or in digital form using a transient
recorder. After the test the information may be copied to
a computer and may be analysed numerically.

1.3 Uniaxial loading

Axial loading is often assumed to be the test which is the
physically most sound because the displacement distribution
is equal over the cross-section. This is true for a
homogeneous isotropic material. It becomes doubtful when
cracks occur. In a plastic hardening material, equilibrium
will force the uniaxial specimen into a centric position
when it starts to become eccentric. However, a specimen of
an elastic softening material tends to become more
eccentric because areas of softening carry less and less
load (Hillerborg 1989). This means that the descending
branch of the stress deformation line may be erroneous if
too long a specimen is used which is able to rotate even if
the specimen ends do not rotate.

With this in mind, reliable tests can be performed on
short specimens which are loaded by a servo hydraulic
actuator, by drop weights like in the split-Hopkinson bar,
or by stress waves.

1.3.1 Inertia effect

Let us assume a tensile specimen which is loaded at the
lower end according to Fig. 10. The vertical displacement
and acceleration distribution are linear over the specimen
length l. The inertia resistance then is

$$P_1(t) \, \delta u_0 = \int_0^l \varrho A \ddot{u}_0 \, \delta u_0 \, \frac{x^2}{l^2} \, dx \qquad (7)$$

With ϱ = const and A = const

$$P_1(t) = \varrho A \ddot{u}_0 \, \frac{l}{3} \qquad (8)$$

This means that for constant displacement velocity there
is no inertia effect and that acceleration and length of
the specimen are directly proportional to the inertia
effect.

If we assume that a cracking zone starts to develop in
the centre of the specimen, the inertia force becomes (Fig.
11)

$$F_1(t) = \varrho A \ddot{u}_0 \; \frac{1}{2} \tag{9}$$

which is more than in the previous case.

An example can illustrate whether inertia forces are very relevant in a uniaxial test. If we assume a specimen of 100 mm length, a maximum crack opening of 200 μm at failure, which is reached after 1 ms, we get for normal weight concrete with $\varrho = 2400$ kg/m³ a mean acceleration of $\ddot{u}_0 = 400$ m/s² and an inertia stress

$$\frac{P_1(t)}{A} = 2400 \cdot 400 \cdot \frac{0.1}{2} = 48,000 \; Pa = 0.048 \; MPa \tag{9a}$$

Compared to a tensile strength of 4.8 MPa, the inertia influence is only 1 %.

1.3.2 Strain and deformation measurement
In high rate loading, test strain gauges can be used, LDVTs and proximity transducers. They should fulfil the requirements with respect to the frequency of the signal to be measured. The frequencies can reach about 10 kHz.

1.3.3 Force measurement
It is usual to measure the force by a load cell which is placed in series with the specimen. In order to minimize inertia effects, it is preferable to mount the load cell directly at the non—loaded end of the specimen.

In the split—Hopkinson bar, the transmitter bar is used as a force measuring device (Reinhardt, Körmeling and Zielinski 1986) if rather long loading pulses are applied. In the original split—Hopkinson pressure bar, where a short rectangular pulse is generated, the stress in the specimen can be calculated from the reflected wave, which travels backwards in the incident bar, see Fig. 12. For details of the analysis, the reader is referred to the book of Zukas et al. (1982).

1.3.4 Loading devices
There are hydraulic actuators which can reach a piston velocity up to 20 m/s. These are rather expensive devices and only a few laboratories have them in their repertoire. On the other hand, these actuators also need a certain displacement to develop the full velocity and since concrete is rather brittle, this necessary displacement may be even larger than the fracture displacement of concrete in tension.

Usually, hydraulic equipment is able to reach lower displacement rates, which enable strain rates up to 0.2 s⁻¹ (Paulmann and Steinert 1982). Extra pressure storage can increase the straining rate considerably as has been experienced by Takeda (1959) and others.

Higher strain rates can be achieved by electromagnetic loading devices (Zhurkov 1965). The split—Hopkinson bar has already been mentioned which also allows higher loading rates. The same is true for drop weight loading or blast loading.

An interesting method has been proposed by Gran (1986), who uses two accelerated masses hitting a circular cylinder on both ends simultaneously. The compressive waves are reflected at the remote ends, convert to tensile pulses which meet in the centre of the specimen where they cause fracture. The strain rate at the centre was recorded as large as 20 s^{-1}.

A general review on impact loading has been presented by Reinhardt (1982).

1.4 Other types of loading

There are other types of loading such as the wedge splitting test (Brühwiler and Wittmann 1987), the CT—specimen, the splitting test, the point load test, and maybe others. However, all of them need loading equipment which is basically compressive or tensile. They also need measuring and recording devices which are essentially the same as already described. Therefore it is felt at the moment that those tests do not need additional description.

1.5 Results from testing and theory
1.5.1 Tensile strength

Most research has been devoted to the tensile strength of concrete since this is the quantity which limits the tensile capacity according to the strength of materials approach. A great number of tests have been carried out which show a significant influence of loading rate on the tensile strength. Furthermore, theories have been developed in order to explain the strength increase due to high loading rates on thermodynamic or statistical grounds. Inertia forces in the vicinity of a running crack have also been taken into account, as well as the influence of a limited crack propagation velocity on the behaviour. Before experimental results will be given, several theories will be shortly presented.

1.5.1.1 Fracture theories which include a rate effect

Under the assumption that linear elastic fracture mechanics (LEFM) is valid, Kipp, Grady and Chen (1980) extended the theory of constant stress to arbitrary stress loading by an appropriate use of the stress—time relation. From the special case of a constant strain—rate $\dot{\varepsilon}_0$ and thus a constant stress rate $\dot{\sigma}_0$ in an elastic material, the following relation for the stress intensity factor for a penny—shaped crack is derived

$$K_I(t) = \frac{4\,\alpha}{3\sqrt{\pi}}\,\dot{\sigma}_0\,\sqrt{(c_s)}\,t^{3/2} \tag{10}$$

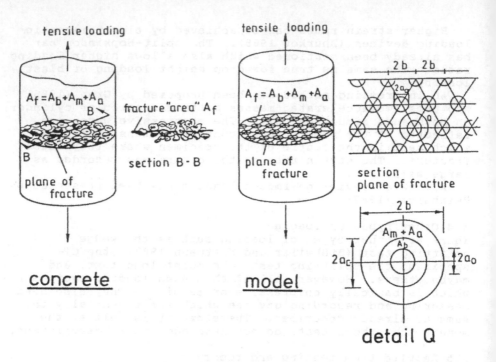

Fig. 13 Voids in concrete represented by penny—shaped cracks

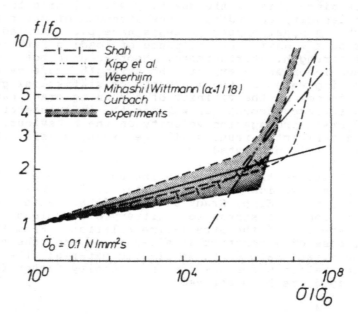

Fig. 14 Comparison of model predictions and experiments with respect to tensile strength of concrete

where α is a geometric coefficient equal to 1.12 for the penny—shaped crack, c_s the shear wave velocity, and t loading time. If K_{Ic} is regarded as a fracture criterion, a relation between strain rate $\dot{\varepsilon}_0$ and strength f (critical stress) can be established

$$f = \left(\frac{9\pi E K_{Ic}^2}{16\alpha^2 c_s} \right)^{1/3} \dot{\varepsilon}_0^{1/3} \tag{11}$$

This cube root law holds for high strain rate and/or sufficiently large cracks.

Weerheijm (1984) used also LEFM but applied it to concrete on a micro level. Concrete is schematized as a material containing penny—shaped cracks of single size and equal distance. Fig. 13 shows a concrete representation element with flaw diameter 2a and centre distance 2b. At the beginning the ratio a/b is calculated from the pore volume n of the concrete; n accounts for the total pore volume not distinguishing between gel pores, capillary pores, and initial shrinkage cracks. Assuming spherical pores of radius a in a fictitious sphere with radius b, the ratio a/b becomes $n^{1/3}$. The absolute values of a and b are determined from the critical stress intensity factor which is taken from macroscopic experiments and a uniaxial tensile strength of 0.6 times the static tensile strength f_t. This means that this stress level is assumed to be a critical stress level where unstable crack propagation starts. These assumptions lead to an expression for a

$$a = \frac{1}{\pi} \left(\frac{K_{Ic}}{0.6 \, f_t \, f(a/b)} \right)^2 \tag{12}$$

where f(a/b) is a geometric function for the regularly flawed material.

The dynamic aspect is treated by considering the kinetic energy during crack propagation. Weerheijm has shown that total energy G_I and kinetic energy E_{kin} depend upon stress, stress rate, initial crack length, crack velocity, Young's modulus and Poisson's ratio.

From the condition

$$E_{kin} = \int_{a_1}^{a_2} (G_I - G_{Ic}) \, da \tag{13}$$

(a_1 and a_2 are two states of cracking) a relation follows between stress and fracture time. For constant stress rate, the tensile strength can be calculated. As a matter

213

of fact, this model is an appropriate application of Mott's (1948) idea to concrete as an inhomogeneous precracked material.

In continuation with this approach, Weerheijm (1989) has stated that the partition between the various types of energy changes during crack extension. This causes a decreasing crack propagation velocity resulting in a strength increase. The available energy to form new crack surfaces in the region around the crack tip decreases.

This description is identical with the observation of Curbach (1987) that the stress distribution in front of microcracks changes due to increasing strain rate. The stress concentration observable at quasistatic loading changes to a more uniform stress distribution. For the same load level this means a transfer of stresses from the region around the crack tip to remote regions. Assuming the same critical strength in front of the crack tip, the more uniform stress distribution leads to the same strength increase as described by Weerheijm (1989).

Often researchers treated the rate influence on thermodynamic grounds, i.e. on an atomic level. Atoms are in a state of continuous motion, attracting and repulsing forces are acting on them. Each atom is situated on a certain energy level and possesses a certain chance to overcome the inherent energy barrier and to move to another place. If external energy is added, the chance to move to one direction becomes larger than to move to the opposite direction.

Mihashi and Wittmann (1980) used this approach to predict the influence of loading rate on strength of concrete. They state that fracture is caused by a series of local failure processes in the hydrated cement paste and in the interfaces between cement and aggregate. As soon as a failure criterion is reached, a crack is initiated at a material defect. The distribution of defects and characteristic properties is statistically equal over the whole material.

The rate of crack initiation is expressed by

$$r = \frac{kT}{h} \exp\left(-\frac{U_0}{kT}\right) (q\sigma)^{\frac{1}{n_b kT}} \tag{14}$$

with k = Boltzmann constant, h = Planck constant, T = absolute temperature, U_0 = activation energy, q = local stress concentration factor, and n_b = a material constant. Eq. (14) becomes rather simple if all other quantities are taken constant

$$r \sim \sigma^\alpha \tag{15}$$

which leads to

$$f/f_0 = (\dot{\sigma}/\dot{\sigma}_0)^\alpha \tag{16}$$

with f and f_0 tensile strength under impact and static loading, and $\dot{\sigma}$ and $\dot{\sigma}_0$ the inherent stress rates. The coefficient α depends on the material, temperature and humidity.

Rate theory has been applied to rock and other materials by Zhurkov (1965) and Lindholm and Nagy (1974). They assume the activation energy to be a linear function of stress

$$U(\sigma) = U_0 - a(\sigma - \sigma_0) \tag{17}$$

with U_0 = total activation energy of the process, a the activation volume, σ_0 a constant, and σ the stress applied.

The rate equation is used in its simplest form

$$\dot{\varepsilon} = \dot{\varepsilon}_0 \exp\left(-\frac{U(\sigma)}{RT}\right) \tag{18}$$

with $\dot{\varepsilon}$ = strain rate, $\dot{\varepsilon}_0$ a constant, R = gas constant, and T = absolute temperature. Eqs. (17) and (18) lead to

$$\sigma_{max} = f = \frac{U_0}{a} + \sigma_0 - \frac{RT}{a} \ln \frac{\dot{\varepsilon}_0}{\dot{\varepsilon}} \tag{19}$$

In this relation, $(U_0/a + \sigma_0)$ is the limiting stress when T = o or when $\dot{\varepsilon} = \dot{\varepsilon}_0$. Eq. (19) predicts that the tensile strength decreases linearly with temperature and increases linearly with the logarithm of the imposed strain rate.

Based on a damage model of Krajcinovic and Fonseka (1981), Shah (1983) used the variation of the Helmholtz free energy function for the formulation of a constitutive equation which describes the behaviour of concrete under varying rates of loading. He showed that tensile strength increases more than linearly with the logarithm of strain rate.

In a recent study, Curbach (1987) showed the very important influence of finite crack propagation velocity on the tensile strength of concrete. By use of a finite element code with appropriate formulation of inertia effects it was demonstrated that strain rates beyond 1 s⁻¹ lead to an increase of tensile strength. At ε = 100 s⁻¹ the predicted strength is six times the static strength and 60 % of the increase is due to crack velocity influence.

Looking on the inhomogeneous structures of concrete, there is a secondary effect which depends on inertia: it is the exact path which is forced through aggregate particles rather than around the particles. Since strength and fracture energy of the aggregate particles are larger than those of the matrix, this effect increases the total fracture energy of concrete and the tensile strength as well. This has been shown by Zielinski and Reinhardt (1981) and has been supported by Brühwiler (1987).

1.5.1.2 Experimental results
Experimental results confirm the theoretical predictions as shown by Fig. 14. It can be seen that some models are appropriate for small to intermediate stress rates whereas other models predict the high stress rates best: The first category is based on rate theory and free energy considerations whereas the second uses inertia effects as a basis.

The left hand part of the diagram is best described by the power law

$$f/f_0 = (\dot\sigma/\dot\sigma_0)^\alpha \tag{16}$$

The power α depends on composition of concrete, humidity and temperature. In a recent investigation by Brühwiler (1987), normal weight concrete with 32 mm maximum aggregate size has been compared with dam concrete with 80 mm maximum aggregate size. It turned out that eq. (16) was supported by both concretes but that the value of α was 0.064 for the first concrete and 0.080 for the dam concrete. This means that the dam concrete appeared to be more sensitive to loading rate.

Opposite to the power function relation, Ross and Thompson (1988) have fitted experimental results by a linear logarithmic function and found reasonable agreement. This function is in accordance with the prediction by eq. (19). It should be noted that scatter of results is considerable and that various functions lead to acceptable agreement.

The effect of compressive preloading on the tensile strength at higher strain rates has been investigated by Brühwiler (1987b). A maximum aggregate size of 32 and 80 mm has been used in two test series. Single compressive preloading up to 750 microstrain and repeated compressive preloading up to 360 microstrain caused a decrease of tensile strength by 40 % on normal aggregate concrete and almost no reduction on large size aggregate concrete. The decrease of strength is more pronounced at high strain rates than it is at low strain rates. The power law (eq. 16) has also been supported in case of preloaded concrete.

1.5.2 Young's modulus
Concrete can be treated as a linear elastic material up to a stress which is about 60 % of the tensile strength. The attainable strain is rather small (about $50 \cdot 10^{-6}$) and there are only micro—flaws in the material which extend a little, i. e. not a large area around a running crack which contributes to the resistance due to inertia effects. Therefore it is expected that Young's modulus is less affected by stress rate than strength is or even not affected at all (as Shah reports 1987). From experiments

the following tentative relation can be derived which does not seem to depend on concrete grade

$$E/E_0 = (\dot\sigma/\dot\sigma_0)^{0.016} \tag{20}$$

with $\dot\sigma_0 = 0.1$ MPa/s.

It should be noted that eq. (20) is only valid for $\dot\sigma/\dot\sigma_0 > 1$. At slow loading rates or even at constant stress, creep comes into play which may cause an apparent elastic modulus which is much smaller than would have been predicted by eq. (20). However, creep and elastic strain should be treated appropriately in order to prevent confusion.

1.5.3 Ultimate strain
"Ultimate" strain is an expression which stems from the time when testing was performed force—controlled. At peak stress, failure occurred and the strain measured at that moment was called ultimate. Meanwhile it is well known that displacement continues with decreasing stress. However, ultimate strain is still meaningful because it determines the transition point between continuum behaviour and discrete cracking behaviour.

The stress—strain curve of concrete in tension first shows a straight line (given by Young's modulus) and later a curved line which is due to progressive cracking. Shah's model (1983) predicts an increase of ultimate strain with increasing strain rate. Other models apply only to strength and do not predict deformation behaviour.

From experiments an average relation between ultimate strain and stress rate can be established which is

$$\varepsilon_u/\varepsilon_{u.o} = (\dot\sigma/\dot\sigma_0)^{0.02} \tag{21}$$

The index o refers to static loading with $\dot\sigma_0 = 0.1$ MPa/s. Up to ε_u, stress rate may approximately be converted to strain rate by $\dot\varepsilon = \dot\sigma/E$.

1.5.4 Traction free crack opening displacement
The elongation of a plain concrete specimen under uniaxial tensile force is composed of elastic strain times specimen length and opening of a discrete crack. This is correct if one follows the assumption of the discrete crack model. Fig. 15 shows the decomposition of total displacement into these two parts. δ_0 is the displacement where no stress is transferred any more.

There is little information available with respect to an accepted traction free crack opening displacement as such and even less with respect to the influence of rate of displacement on δ_0. There are authors who found an increase, others found a constant value at high displacement rates. For the time being, it is proposed to hold on a value of δ_0 which is rate independent.

217

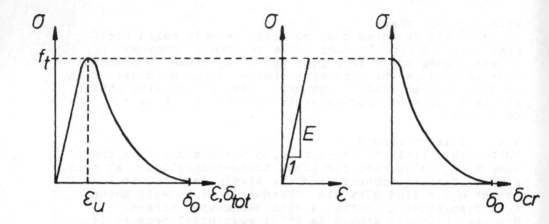

Fig. 15 Decomposition of total displacement into elastic
strain and crack opening displacement

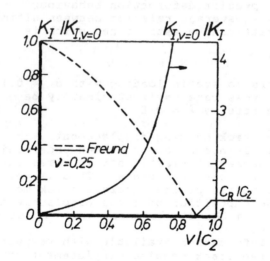

Fig. 16 Semi—infinite crack propagating in a finite strip

1.5.5 Fracture energy

Fracture energy G_F is the area under the $\sigma-\delta_{cr}$ curve of Fig. 11. If we assume δ_0 = const, then G_F depends on tensile strength and the shape of the descending branch. If tensile strength increases with stress rate and the shape of the descending branch remains similar, then G_F increases similarly to f_t. If the shape of the descending branch changes, then G_F shows another relation to displacement rate than f_t does.

There are not many experimental results available which show a clear rate dependent behaviour. Brühwiler and Wittmann (1987) suggest describing the rate influence by two terms, the first of which accounts for a creep influence at small rates while the second describes the high rates:

$$G_F/G_{F0} = a \; (\dot{\delta}_0/\dot{\delta})^\beta + b \; (\dot{\delta}/\dot{\delta}_0)^\alpha \qquad (22)$$

$a \approx 0.15$, $b \approx 0,85$, β and α depend on concrete mix, humidity and temperature. Typical values are: $0.3 < \beta < 0.45$ and $0.075 < \alpha < 0.125$.

Reinhardt (1987) evaluated attainable results and came to a single power law which reads

$$G_F/G_{F0} = (\dot{\delta}/\dot{\delta}_0) \qquad (23)$$

with $\gamma = (10 + f_{cm}/2)^{-1}$. Here f_{cm} is the mean cube compressive strength of concrete in MPa and $\dot{\delta}_0 = 1 \; \mu m/s$. For usual concretes, γ becomes 0.025 to 0.05.

Thermally activated processes can be described by rate theory which has been applied to concrete by Körmeling (1986). The following relation was used

$$G_F = a + f(T) \; (\ln \dot{\delta} - b) \qquad (24)$$

where a is the macroscopic activation energy per cross-section, f(T) is a function of absolute temperature and b a constant. For a certain normal concrete eq. (24) becomes

$$G_F = 565 + 0.37 \; T(\ln\dot{\delta} - 10) \qquad (25)$$

with $\dot{\delta}$ in mm/s, G_F in J/m^2, T in K. So G_F depends linearly on absolute temperature and logarithm of displacement rate. It should be noted that eq. (25) is only valid in a limited range in which the thermally activated process is secured. More research is needed in this area to establish a clear picture.

1.5.6 Fracture toughness

Fracture toughness K_{Ic} characterizes the brittleness of an elastic cracked material. For a certain given crack length, a low K_{Ic} means unstable crack extension at low remote stresses whereas at high K_{Ic} a high stress is necessary. On the other hand, the stress intensity factor K_I characterizes the magnitude of stress around a crack tip. Thus K_{Ic} is a material property while K_I is a scaling function.

There are several researchers who derived stress intensity factors for moving cracks. Freund (1972) for instance determined K_I for a semi—infinite crack in a finite strip. At time t = 0 the faces of the strip are moved by an instantaneous displacement which is kept constant. Fig. 12 shows the result. It appears that K_I decreases with crack propagation velocity v. At v = c_R (Rayleigh speed) K_I = 0. If K_{Ic} is a material property, this means that the apparent K_{Ic}, i. e. when unstable crack extension starts, increases with higher rate of loading. This is depicted by the solid line in Fig. 16.

If crack velocity v is known, Fig. 12 can be evaluated. Shah and John (1986) show a linear relationship between logarithm of crack velocity and logarithm of strain rate. At ε = 1 s^{-1} velocity becomes about 200 m/s. With shear wave speed $c_2 \approx$ 2600 m/s v/c_2 = 0.12 and thus becomes the apparent fracture toughness (Fig. 16) about 10 % larger than in the static case. Muria Vila and Hamelin (1987) carried out tests with high strain rates and report on crack velocities up to 1200 m/s. In that case v/c_2 = 0.46 and thus K_{Ic} = 1.7 $K_{Ic,0}$. These figures make evident that a tremendous increase of K_{Ic} would exist only at crack velocities which have not yet been measured on concrete.

Wittmann (1987) measured K_{Ic} on autoclaved aerated concrete and found a power law between K_{Ic} and deflection rate

$$K_{Ic}/K_{Ic,0} = (\dot{d}/\dot{d}_0)^{0.092} \tag{26}$$

with $\dot{d}_0$ = 17 µm/s.

It should be noted that fracture toughness stems from LEFM which demands large specimens to be used. It is therefore that reliable experimental results are still missing. Rossi (1986) used large double cantilever beams but could not support eq. (26); he found even a small decrease of K_{Ic} with increasing loading rate. However, the loading rate was so small that viscoelastic effects may have played a role.

1.5.7 Crack propagation velocity

The crack propagation velocity in concrete has a considerable influence on the behaviour of concrete under high strain rates, as it has been shown by Weerheijm (1984). The crack velocity in linear elastic materials is limited to the Rayleigh—wave velocity but it has been observed that almost none of the materials have reached this velocity, see e.g. Broek (1984).

Not many results are available on the crack velocity in concrete. Curbach, Hehn and Eibl (1988, 1989a) have carried out tests on the crack velocity in concrete using crack gauges consisting of liquid silver barriers which were destroyed by the propagating crack.

The tests indicate that the maximum value of crack propagation velocity of an unstable crack is independent of the strain rate. A value of about

$$V_{crack} \approx 500 \; m/s$$

has been found which is less than a quarter of the theoretical maximum value of about 2100 m/s which is about the Rayleigh-wave velocity of concrete.

Crack propagation velocity values of about 200 m/s have been reported by Shah et al. (1986) while Muria Vila et al. (1987) report test results from compressive tests on concrete where velocities up to 1200 m/s have been measured. It should be noted, however, that the crack propagation measurement used by Shah was more sensitive than the one used by Muria Vila.

Tests and finite element analyses performed at the University of Washington, Seattle, have shown that the maximum propagation velocity of a running crack in a bend specimen amounts to about 300 m/s if the specimen is loaded by a falling hammer. A smaller crack propagation velocity of about 130 m/s has been determined in a bending test with a maximum strain rate of 0.1 s^{-1} (Kobayashi, Hawkins and Du, 1989, and Yon, Hawkins and Kobayashi, 1989b).

1.5.8 Two-parameter model

John and Shah (1987) extended the original two-parameter model which was derived for static loading (Jenq and Shah 1985), to rate effects. The model itself is described in another subcommittee's report (Subcommittee A: Notched beam test: Mode I Fracture Toughness). Here, only the rate influence will be discussed. The authors use $K_{Ic}{}^s$ which is an effective fracture toughness value and $CTOD_c$ which is the critical elastic crack tip opening displacement, both of which are determined in a deflection controlled bending test. To use two parameters instead of one has the advantage that the real crack length has not to be determined accurately. This is always a difficulty since pre-critical crack growth cannot be clearly measured at the surface.

The authors assume $K_{Ic}{}^s$ to be rate independent since crack velocities are small and the rate influence on $K_{Ic}{}^s$ should also be small (compare Fig. 12). On the other hand, $CTOD_c$ is assumed to be rate dependent with the following relation

$$\frac{CTOD_c}{CTOD_{c,o}} = \exp \left[-\frac{A_1}{\sqrt{f_c}} \left(\log \frac{\dot{\varepsilon}}{\dot{\varepsilon}_0} \right)^B \right] \qquad (27)$$

with index o for static loading, $\dot{\varepsilon}_0 = 0.1 \cdot 10^{-6}$ s^{-1}, f_c = mean concrete cylinder compressive strength. In an example

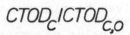

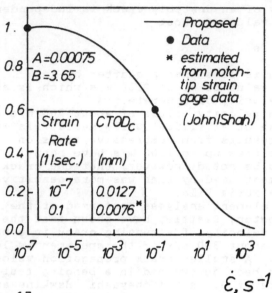

Fig. 17
CTODc as a function
of strain rate

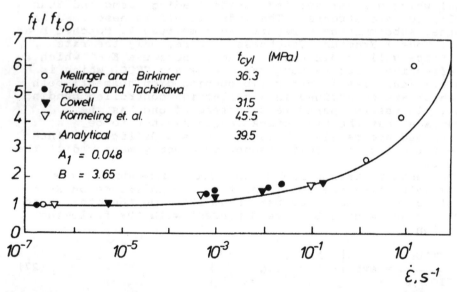

Fig. 18   Predicted relation between tensile strength and
strain rate according to two—parameter model

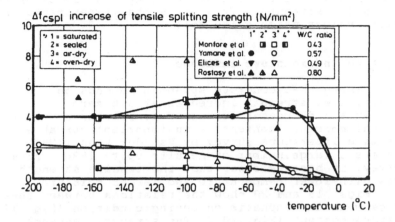

Fig. 19 Tensile splitting strength vs. curing condition and temperature

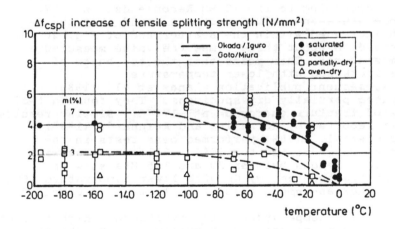

Fig. 20 Tensile splitting strength vs. moisture content and temperature

$A_1 = 0.00075$ and $B = 3.65$, $f_c = 27.5$ MPa. Fig. 17 shows eq. (27) and Fig. 18 gives the result of an analysis in which the two—parameter model has been applied in order to predict tensile strength. It appears that experimental and theoretical values match quite well.

2 Temperature and moisture influence

It is well known that concrete strength depends on temperature, i.e. strength increases at sub—zero temperatures and decreases at higher temperatures. In this context, the moisture content is an important variable.

There is a literature survey by Van der Veen (1987) which covers a large number of results up to 1986. From this survey, it becomes clear that the tensile strength increases with lower temperature and larger moisture content. Figs. 19 and 20 show the relations between these quantities with the emphasis on curing condition (Fig. 19) and moisture content (Fig. 20). The strength increase is most pronounced in the temperature range of 0 to −100°C.

Mihashi et al. (1987) report on bending tests which show also an increase of bending strength at freezing temperature, however, the dry specimens are stronger than the wet specimens. This effect is attributed to microcracking during freezing.

Elices et al. (1987) performed tests on water saturated beams and determined the tensile strength f_t and fracture energy G_F according to the RILEM Recommendation (1974) at room temperature and −170°C. It was found that the tensile strength increased within the scatterband of Fig. 20 while G_F increased to about three times the value measured at room temperature. This means that the toughness of concrete increased with lower temperature.

In a subsequent publication Planas et al. (1989) investigated partially dry specimens. They found a twofold increase of fracture energy G_F at −170°C. Similar results have been reported by Körmeling and Reinhardt (1987).

Ohlsson et al. (1988) performed beam tests in the temperature range of +20 to −35°C. All concretes with compressive strength between 30 and 105 MPa show an increasing fracture energy G_F with lower temperature (Fig. 21). The increase amounts about 50 % at −35°C compared to +20°C.

Tests at low temperature with simultaneous high rate of loading have been reported by Körmeling and Reinhardt (1987). Although the number of tests was limited, a relation between rate of displacement, temperature, and fracture energy could be established:

$$G_F = 565 − 0.37\ T\ (10 − \ln \dot{\delta}) \tag{28}$$

with T = absolute temperature in K, $\dot{\delta}$ = displacement rate in mm/s of a measuring length of 35 mm, and G_F = fracture

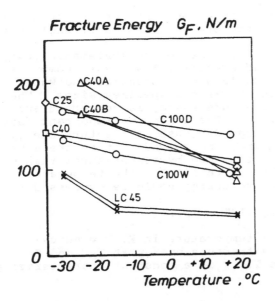

Fig. 21 Fracture energy vs. temperature for various concrete mixes

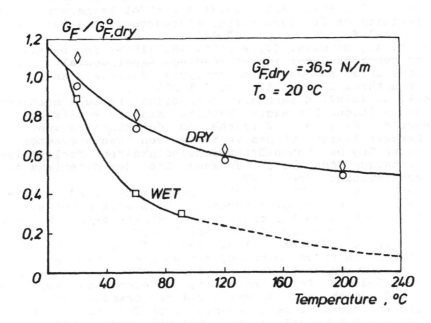

Fig. 22 Fracture energy at elevated temperature

energy in J/m^2. The three constants depend on concrete grade. The actual concrete compressive strength was 44 MPa.

Bazant and Prat (1988) performed systematic tests on bending specimens and eccentrically loaded compressive specimens at +20°C to +200°C in predried and water saturated condition at various specimen sizes. The main results as far as fracture energy is concerned are summarized by Fig. 22. There are two important features: first, there is a large decrease of fracture energy with increasing temperature, and second, wet concrete is considerably more influenced by temperatures than dry concrete, especially up to 100°C. The temperature dependence of fracture energy could be expressed by

$$G_F = G_{F.o} \exp [\gamma (1/T - 1/T_o)] \tag{29}$$

with T_o = reference temperature in K, T = actual temperature, $G_{F.o}$ = fracture energy at T_o, and G_F = fracture energy at T. γ is a constant characterizing the given material.

3 References

Bazant, Z.P., Prat, P.C. (1988) Effect of temperature and humidity on fracture energy of concrete. ACI Materials Journal 85, no. 4, pp 262–271

Bentur, A., Mindess, S., Banthia, N. (1986) The behaviour of concrete under impact loading: Experimental procedures and method of analysis. Materials and Structures 19, no. 113, pp 371–378

Broek, D. (1984) Elementary engineering fracture mechanics, 3rd edition. The Hague: Martinus Nijhoff Publishers

Brühwiler, E. (1987a) Versuche über den Einfluss von Druckvorlasten auf das Verhalten von Staumauerbeton unter Zug bei hohen Dehngeschwindigkeiten. Ecole Poly-technique Fédérale de Lausanne, Lab. des Matériaux de construction, Dec. 1987

Brühwiler, E. (1987b) Effect of compressive loads on the tensile strength of concrete at high strain rates. Contribution to the draft of the state–of–the–art report of RILEM Committee 90–FMA, April 1987

Brühwiler, E., Wittmann, F.H. (1987) Effect of rate of loading on fracture energy and strain softening. Contribution to RILEM Committee 89–FMT. Stevin report no. 25–87–16, Delft University of Technology, April 1987

Bur, A.J., Roth, S.C. (1985) A polymer pressure gage for dynamic pressure measurements. Proc 2nd Symp. Inter-action of non–nuclear munitions with structures. Panama City Beach, Florida, April, pp 291–295

Chung, R.M., Bur, A.J., Holder, J.R. (1985) Laboratory evaluation of an NBS polymer soil stress gauge. Proc. 2nd Symp. Interaction of non—nuclear munitions with structures. Panama City Beach, Florida, April, pp 296—301

Curbach, M. (1987) Festigkeitssteigerung von Beton bei hohen Belastungsgeschwindigkeiten. Schriftenreihe des Instituts für Massivbau und Baustofftechnologie, Heft 1, Karlsruhe, 154 pp

Curbach, M., Eibl, J. (1988) Crack propagation in concrete. Conference Fracture of Concrete and Rock, 1988, to be published

Curbach, M., Hehn, K.—H., Eibl, J. (1989a) Measurement of crack velocity in concrete, Experimental Techniques 13 (1989), pp. 25—27

Curbach, M., Maliszkiewicz, P., Eibl, J. (1989b) Acoustic emission measurement of concrete under high loading rates. Transactions of the 10th International Conference on Structural Mechanics in Reactor Technology SMiRT

Diederichs, U., Schneider, U., Terrien, M. (1983) Formation and propagation of cracks and acoustic emission, Fracture Mechanics of Concrete, Ed. F.H. Wittmann, Amsterdam: Elsevier

Elices, M., Planas, J. (1982) Measurement of tensile strength of concrete at very low temperatures. ACI J. 79, no. 3, pp 195—200

Elices, M., Planas, J., Maturana, P. (1987) Fracture of concrete at cryogenic temperatures. SEM/RILEM Intern. Conf. Fracture of Concrete and Rock, ed. S.P. Shah and S.E. Swartz, Houston, June 1987, pp 159—169

Frank, T. (1984) Beeinflussung der Prüfergebnisse durch die Messeinrichtung bei der Stossbeanspruchung von Beton. Materialprüfung 26, no. 4, pp 96—100

Freund, L.B. (1972) Crack propagation in an elastic solid subjected to general loading — I. Constant rate of extension. J. Mech. Phys. Solids 20, pp 129—140

Gopalaratnam, V.S., Shah, S.P., John, R. (1984) A modified instrumented Charpy test for cement—based composites. Experimental Mechanics 24, no. 3, pp. 102—111

Goto, Y., Miura, T. (1978) Mechanical properties of concrete at very low temperatures. Proc. 21st Japanese Congress on Material Research, Kyoto, pp 157—159

Gran, J.K. (1986) A new technique for studying the dynamic tensile response of concrete. Proc. 2nd Symp. Interaction of non—nuclear munitions with structures. Panama City Beach, Florida, pp 206—210

Hillerborg, A. (1989) Stability problems in fracture mechanics testing. In "Fracture of concrete and rock: Recent developments." eds. S.P. Shah, S.E. Swartz, B. Barr. Elsevier Appl. Sci. London, New York, pp 369—378

Hordijk, D.A., Reinhardt, H.W., Cornelissen, H.A.W. (1987) Fracture mechanics parameters of concrete from uniaxial tensile tests as influenced by specimen length. Proc. Int. Conf. on Fracture Mechanics of Concrete and Rock, Houston (USA), pp 138—149

Hughes, G., Beeby, A.W. (1982) Investigation of the effect of impact loading on concrete beams. The Structural Engineer 60 B, no. 3, pp 45—52

Jenq, Y.S., Shah, S.P. (1985) A two parameter fracture model for concrete. J. Eng. Mech. ASCE 111, No. 10, pp 1227—1241

John, R., Shah, S.P. (1987) Effect of high strength and rate of loading on fracture parameters of concrete. Proc. RILEM—SEM Intern. Conf. Fracture of concrete and rock. Houston, eds. S.P. Shah and S.E. Swartz, pp 35—52

John, R., Shah, S.P. (1986) Fracture of concrete subjected to impact loading. ASTM Cement, Concrete and Aggregates 8, pp 24—32

Kipp, M.E., Grady, D.E., Chen, E.P. (1980) Strain—rate dependent fracture initiation. Int. J. Fracture 16, pp 471—478

Kobayashi, A.S., Hawkins, N.M., Du, J.J. (1989) An impact damage model of concrete. University of Washington, Seattle

Körmeling, H.A. (1986) Strain rate and temperature behaviour of steel fibre concrete in tension. Doctoral thesis, Delft University of Technology

Körmeling, H.A., Reinhardt, H.W. (1987) Strain rate effects on steel fibre concrete in uniaxial tension. Intern. J. Cement Composites and Lightweight Concrete 9, no. 4, pp 197—204

Lindholm, U.S., Yeakley, L.M., Nagy,A. (1974). The dynamic strength and fracture properties of Dresser basalt. Int. J. Rock Mech. Min. Sei. and Geomech. Abstracts 11, pp 181—191

Mihashi, H., Nomura, N., Kinoshita, Y. (1987) Constitutive law with strain softening of freezing mortar. Annual Report of Cement Ass. of Japan, pp 1—2

Mihashi, H., Wittmann, F.H. (1980) Stochastic approach to study the influence of rate of loading on strength of concrete. Heron 25, Nr. 3

Mindess, S., Banthia, N.P., Ritter, A., Skalny, J.P. (1986) Crack development in cementitious materials under impact loading. eds. S. Mindess and S.P. Shah. MRS Symp. Proc. Voc. 64, Pittburgh, PA, pp 217—223

Monforce, G.E., Lentz, A.E. (1962) Physical properties of concrete at very low temperatures. J. PCA, Research and Development Laboratories, pp 33—39

Mott, N.F. (1948) Fracture of metals: Theoretical considerations. Engineering 165, pp 16—18

Muria Vila, D., Hamelin, P. (1987) Comportement au choc des bétons et mortiers à matrices hydrauliques. 1st Int. RILEM Congress Vol. 2 "Combining materials: Design, production and properties", ed. J.C. Maso, Chapman and Hall, London New York, pp 725—732

Ohlsson, U., Daerga, P.A., Elfgren, L. (1988) Fracture energy and fatigue strength of unreinforced concrete beams at normal and low temperatures. Preprint "Intern. Conf. Fracture and Damage of Concrete and Rock", Vienna, July 1988

Okada, T., Igur, M. (1978) Bending behaviour of prestressed concrete beams under low temperature. J. Japanese Pre—stressed Concrete Ery. Ass. Special Issue for 8th FIP Congress

Paulmann, K., Steinert, J. (1982) Beton bei sehr kurzer Belastungsgeschichte. Beton 32, no. 6, pp 225—228

Planas, J., Maturana, P., Guinea, G., Elices, M. (1989) Fracture energy of water saturated and partially dry concrete at room and at cryogenic temperatures. Intern. Congress of Fracture, Houston, March 1989, pp 1—9

Reinhardt, H.W. (1982) Testing and monitoring techniques for impact and impulsive loading of concrete structures. Proc. Intern. RILEM Symp. Impact and impulsive loading of concrete structures, Berlin, Vol. I, pp 65—88

Reinhardt, H.W., Körmeling, H.A., Zielinski, A.J. (1986) The split Hopkinson bar, a versatile tool for the impact testing of concrete. Materials and Structures 19, no. 109, pp 55—63

Reinhardt, H.W. (1987) Simple relations for the strain rate influence on concrete. Darmstadt Concrete 2, pp 203—211

Ross, C.A., Thompson, P.Y. (1988) High strain—rate tensile tests of concrete and mortar. Intern. Conf. on Fracture and Damage of concrete and rock, Vienna, Austria, paper no. IX—2

Rossi, P. (1986) Fissuration du béton: Du materiau à la structure. Application de la méchanique linéaire de la rupture. Doctoral thesis ENPC, Paris Dec. 1986

Rossi, P., Le Maour, F. (1986) Ouvrages d'art: le contrôle des fissurations. La Revue des Laboratoires d'essais, no. 6, pp. 25—27

Rostásy, F.S., Scheuermann, J. (1984) Verbund und innerer Zwang von einbetoniertem Bewehrungsstahl bei tiefer Temperatur. Research Report, Braunschweig University of Technology, 92 pp

Shah, S.P. (1983). Constitutive relations of concrete subjected to a varying strain rate. Symp. Proc. The interaction of non—nuclear munitions with structures. US Air Force Academy, Colorado, May 10—13, pp 81—84

Shah, S.P., John, R. (1986) Rate—sensitivity of mode I and mode II fracture of concrete. Mat. Res. Soc. Symp. Proc. Vol. 64, eds. S. Mindess and S.P. Shah, Pittsburgh, pp 21—37

Shah, S.P. (1987) Strain rate effects for concrete and fibre reinforced concrete subjected to impact loading. Final report. Centre for Concrete and Geomaterials. Dept. of Civil Eng. Northwestern University, Evanston, pp 80

Suaris, W., Shah, S.P. (1981) Inertial effects in instrumented impact testing of cementitious composites. ASTM J. Cement, Concrete and Aggregates 3, no. 2, pp. 77–83

Suaris, W., Shah, S.P. (1983) Properties of concrete subjected to impact. ASCE J. Structural Eng. 109, no. 7, pp 1727–1741

Takeda, J. (1959) A loading apparatus for high speed testing of building materials and structures. Proc. 2nd Japan Congress on Testing Materials, Jap. Soc. Testing Mat., Kyoto, Japan, pp 236–238

Van Mier, J.G.M. Pruijssers, A.F., Reinhardt, H.W., Monnier, T. (1990) Load–time response of colliding concrete bodies. Submitted for publication

Van der Veen, C. (1987) Properties of concrete at very low temperature. Stevin Report 25–87–2, Delft University of Technology, Delft, Netherlands, 119 pp.

Weerheijm, J. (1984) Crack model for concrete under dynamic tensile load (in Dutch), Report TNO–PMC 1984–18, The Hague

Weerheijm, J., Reinhardt, H.W. (1989) Modelling of concrete fracture under dynamic tensile loading. Fracture of concrete and rock – Recent developments, Eds. S.P. Shah, S.E. Swartz and B. Barr, London: Elsevier, pp 721–728

Wittmann, F.H. (1987) Influence of rate of loading on the fracture toughness of autoclaved aerated concrete. Contribution to RILEM Committee 89–FMT. Stevin report no. 25–87–16, Delft University of Technology

Yamam, S., Kasami, H., Okuno, T. (1977) Properties of concrete at very low temperatures. ACI Proc. Conf. Mexico City, pp 67–78

Yon, J.H., Hawkins, N.M., Kobayashi, A.S. (1989a) On the strain rate sensitivity of concrete mechanical properties. University of Washington, Seattle, paper (5–15–89)

Yon, J.H., Hawkins, N.M., Kobayashi, A.S. (1989b) Dynamic fracture testing of concrete bend specimens. University of Washington, Seattle, Appendix B

Zhurkov, S.N. (1965) Kinetic concept of the strength of solids. Int. J. Fracture Mechanics 1 (1965), pp 311–323

Zielinski, A.J., Reinhardt, H.W. (1981). Stress–strain behaviour of concrete and mortar at high rates of tensile loading. Cement and Concrete Research 12, pp 309–319

Zukas, J.A. et al. (1982) Impact dynamics. John Wiley & Sons, New York

5 FRACTURE PROCESS ZONE DETECTION

S. MINDESS
University of British Columbia, Vancouver,
Canada

1 Introduction

The propagation of a crack in concrete involves a great deal of microcracking, much of it occurring in the highly stressed region ahead of the apparent crack tip. According to Bazant and Oh (1983), because of the heterogeneity of concrete, it behaves in a nonlinear fashion in a relatively large region adjacent to the apparent fracture front; the fracture process zone represents that part of the nonlinear zone in which the material undergoes progressive microcracking, manifested by strain softening behaviour. On the other hand, Shah (1988) has equated the fracture process zone as the region of precritical crack growth (or slow crack growth) which precedes the maximum load. Thus, while there is general agreement that a fracture process zone exists in concrete, there is no agreement on exactly what constitutes a "process zone" in cementitious materials.

Similarly, there is also no agreement on whether the fracture process zone is a true material parameter. The author of this report has argued elsewhere (Mindess, 1988) that the microcracking referred to above is not necessarily a function of the stress concentrations induced by the crack tip itself, and does not constitute a coherent "fracture process zone" which is a fundamental material parameter. Further, as Thouless (1988) has recently pointed out, instead of looking for a process zone ahead of the crack tip, we can instead consider a bridging zone behind the crack tip. Both approaches yield equivalent results when applied to the problem of crack tip propagation; they differ primarily in the definition of the crack tip.

To further complicate the matter, it has been pointed out by Kasperkiewicz (1988) that there really is no unequivocal definition of a crack length in concrete; once a cast or sawn notch begins to propagate under load, the position of its tip is uncertain. While it is possible to define a "visible" crack and its tip as the simply connected edge of the visible material discontinuity in the topological sense, this will depend on the method of observation. Moreover, because of the heterogeneity of the structure of concrete, there will be discontinuities in the crack path in virgin concrete which act as pre-existing microcracks. It may be necessary to distinguish between these pre-existing flaws and those

created in the vicinity of the crack tip as the crack
begins to grow.

Bazant (1988) has succinctly placed these apparent
discrepancies in the proper perspective, and it is worth
quoting him extensively. In his view,

> if a specimen is sufficiently large, then most of
> the specimen is in an elastic state, and from the
> remote elastic field one can find by optimum fitting
> the location of an elastically equivalent crack tip,
> whose distance from the notch gives the size of the
> fracture process zone, defined in mechanical terms.
> Now, of course, this definition does not have to be
> equivalent to what we see, e.g., under a microscope,
> however it is sufficient for predicting the beha-
> viour of structures of various sizes and shapes
> within a reasonable size range, and that is all that
> we need. It is strictly in this sense that ... the
> fracture process zone is "determined by the size of
> the inhomogeneity in the microstructure", and [this]
> statement should not be misinterpreted by comparing
> it with sizes obtained by direct observations... .
> The point is not to find whether the sizes are
> different, but whether these observations can be
> related. It is of paramount importance ... to
> distinguish between the various definitions of the
> fracture process zone. They are simply different
> animals, and no wonder the sizes are very different.
> In the same sense, the width of the fracture zone in
> the crack band model is strictly the equivalent zone
> which yields the correct energy dissipation and
> correct stress-displacement relation, and cannot be
> compared directly to experiments.

Nonetheless, whatever the "truth" about the existence
of a fracture process zone, it is clearly important to try
to quantify the extent of the damage that occurs during
the fracture of a concrete specimen. Thus, for the purposes of this report, the fracture process zone will be
defined as the region of discontinuous microcracking ahead
of the continuous (visible) crack. However, it must
always be remembered that, as Shah (1988) pointed out, a
substantial traction exists behind the crack tip, which
"can be attributed to frictional [geometrical or aggre-
gate] interlock and to unbroken ligaments".

Finally, it must also be noted that the crack length
appears to vary across the width of a specimen (Swartz and
Go, 1984; Bascoul et al., 1987). Thus, any techniques
used to identify the extent of the process zone which
depend solely on surface crack measurements are likely to
be inadequate for this purpose, even though they may
provide valuable information about the nature of the
microcracking.

The objective of Subcommittee E: Fracture Process Zone Detection, is "to develop experimental methodologies to detect the shape and size of the process [damage] zone at the crack tip". In recent years, a number of experimental studies have been carried out, using a variety of experimental techniques, to measure the extent of the process zone; some of this work is summarized here. This does not purport to be an exhaustive literature review; only a representative sampling of the more recent papers covering the various techniques is provided. Out of this, it is to be hoped that the techniques which seem to hold the most promise for an accurate description of the process zone can be identified.

Whether or not the fracture process zone is a true material parameter, there is general agreement that a zone of discontinuous microcracking does exist ahead of the continuous crack. There is still, however, no agreement on either the extent or the nature of this zone. Estimates of size range from several mm to more than 500 mm. As with other fracture mechanics measurements in cementitious materials, the values obtained seem to depend upon specimen size and geometry, and the technique used to identify the process zone.

The techniques that will be discussed are:
i)   Compliance measurements
ii)   X-ray techniques
iii)   Optical microscopy
iv)   Scanning electron microscopy
v)   Electric resistance strain gauge techniques
vi)   Photoelastic methods
vii)   Mercury penetration measurements
viii)   Dye penetrants
ix)   Infrared vibrothermography
x)   Ultrasonic pulse velocity
xi)   Demec gauges
xii)   Acoustic emission
xiii)   Interferometry techniques
xiv)   Multi-cutting techniques
xv)   Numerical methods

2 Compliance Measurements

Some of the first attempts at estimating the size of the process zone involved compliance measurements. That is, based on measurements of specimen deflections, changes in stiffness, and sometimes the location of the visible crack tip, the extent of the process zone could be "guesstimated". Such determinations are difficult to interpret, since the reduction in stiffness with increased deflection is due in part to slow (subcritical) growth of the continuous crack, and in part to the development of a process zone ahead of the continuous crack. It is generally not possible to separate these two effects.

It is, however, possible to make reasonably good estimates of the fracture process zone. For instance, Karihaloo and Nallathambi (1989) developed the expression for a three-point bend specimen:

$$\frac{a_e}{w} = \gamma_1 \ (\frac{\sigma_n}{E})^{\gamma_2} \ (\frac{a_0}{w})^{\gamma_3} \ (1 + \frac{g}{w})^{\gamma_4} \tag{1}$$

where a_e = effective notch depth, w = beam depth, E = elastic modulus, σ_n = nominal flexural strength, a_0 = initial notch depth, g = maximum aggregate size, and $\gamma_1, \gamma_2, \gamma_3, \gamma_4$ = constants. The size of the fracture process zone is then $a_e - a_0$. Later work (Karihaloo and Nallathambi, 1990) showed that $a_e - a_0$ increased as the specimen depth decreased, ranging from about 40 mm for a beam depth of 400 mm to about 20 mm for a beam depth of 100 mm, starting with an initial a_0/w ratio of 0.5.

Kobayashi et al. (1985) used a combination of compliance measurements and a replica technique to estimate the size of the process zone. The replica technique involved the use of an acetylcellulose film to show the extent of surface crack growth (which seemed to extend 12-25 mm further using this technique than when viewed directly under 10x magnification). The difference between the "total" crack length, as determined from compliance measurements, and the macrocrack length as determined by the replica technique, was then taken to be the size of the process zone. The difficulty here is that compliance measurements are not a very good measure of total crack length, since there is considerable difference between the behaviour of a sawn notch and a "natural" crack. Nonetheless, Kobayashi et al. (1985) found that the process zone continued to grow in size as the crack extended, reaching lengths of 114 mm with no sign of reaching a constant value.

In my view, because of the difficulty in separating the slow crack growth from the development of a zone of discontinuous microcracking ahead of the continuous crack, neither simple compliance measurements, nor the more elaborate analytic techniques used by (Bascoul et al., 1987; Karihaloo and Nallathambi, 1989; Karihaloo and Nallathambi, 1990) can provide a good estimate of the size of the process zone.

3 X-Ray Techniques

X-ray techniques to determine the size of the damage zone have been developed, and used extensively, at Cornell University, for example by Slate (1983a), Slate and Hover (1984), and Najjar and Hover (1988). These techniques show up the fairly large cracks which develop when concrete is loaded, but are not sensitive enough to really define a process zone.

4 Optical Microscopy

Naturally enough, there have been many attempts to observe crack development and the size of the process zone using optical microscopy. The pioneering work in this area at Cornell University, for instance by Slate and Hover (1984), and Slate (1983b), is well known to anyone who has studied the cracking of concrete. The technique employed at Cornell involves the dye impregnation of polished slices of concrete, followed by viewing with a stereo-microscope at magnifications of, typically, 4x to 40x. These relatively low magnifications, while sufficient to reveal macrocracks, are not sufficient to reveal the damage zone.

More sophisticated optical techniques have been more successful in revealing the process zone. Eden and Bailey (1986) used diffuse illumination with a reflected light microscope to enable the observations of cracks of widths down to 0.1 μm. This method made it possible to observe changes occurring in the process zone ahead of the crack tip. They concluded that stable crack growth involves the formation of a process zone which grows to a character-istic size before failure occurs.

Knab _et al_. (1984;1986) developed a fluorescent thin section technique to observe the fracture zone. Their method involved loading the specimen, impregnating with an epoxy containing a fluorescent dye, and viewing under ultraviolet light at relatively low magnifications. They concluded that crack widths down to 2-3 μm could be detec-ted, and that this technique might be used to observe a process zone, though they did not attempt to do so speci-fically. There was some question as to whether the sample preparation techniques themselves induced some cracking.

Stroeven (1988a;1988b) has used fluorescent oils to help delineate cracks in concrete, and has carried out extensive stereological studies on concrete using this technique. The manually copied crack pattern of an axial section of a grooved concrete specimen subjected to direct tension slightly beyond the ultimate load is shown in Fig. 1 (Stroeven, 1988a). It may be seen that there is very extensive cracking. However, a major part of this cracking was already present in virgin specimens, due to shrinkage during curing. With reference to the results of Fig. 1, Stroeven (1988a) found that cracks developed and coalesced in a stochastic way; there is certainly no apparent "process zone". Areas of crack concentrations only became apparent far beyond the peak load, casting

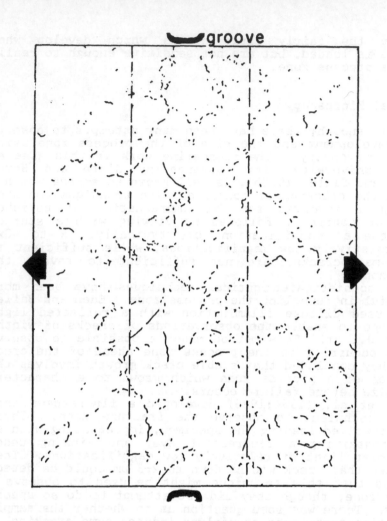

Fig. 1.   Manually copied crack pattern of axial section of
          two-sided grooved specimen subjected to direct
          tension slightly beyond ultimate, after Stroeven
          (1988a).

doubt on the usual assumption of crack growth through the
formation of a fracture process zone in the pre-peak
region of loading.

5 Scanning Electron Microscopy

There has been extensive use of SEM techniques to observe
microcracking in concrete. In principle, because of the
high magnifications which can be used, a process zone, if

one exists, should be easy to observe. However, this has not necessarily been the case; the results vary from investigator to investigator.

For instance, based on extensive in-situ observations of specimens loaded within the sample chamber of an SEM, Mindess and Diamond (1982a;1982b) and Diamond et al. (1983) found very complicated crack patterns, but no evidence of a process zone. Using similar techniques, Diamond and Bentur (1985) found that a system of very fine multiple cracks appeared in the vicinity of the crack tip. They concluded that:

> While the subdivision and branching seen to occur near the tip zone in the concretes examined, is in some respects reminiscent of what is expected in a "process zone", there is no physical distinction corresponding to separate lengths of "straight, open crack" behind a crack tip and "process zone micro-cracking" ahead of a crack tip.

On the other hand, Tait and Garrett (1986), also using in-situ SEM observations, did find a process zone of micro-cracking in the vicinity of the crack tip, 1-4 mm in extent.

Baldie and Pratt (1986) used backscattered electron imaging to examine cracks in polished sections of hardened cement paste. They found "some evidence that crack growth occurs by the formation and coalescence of microcracks ahead of the crack tip, with only a limited formation of an actual process zone". Knab et al. (1986) also used backscattered electron imaging, and were able to view cracks down to a width of 0.5 μm. They did not, however, discuss the existence of a process zone.

Ringot et al. (1987) used a replica technique in conjunction with an SEM, by observing the imprint of the concrete surface on a material insensitive to the environment in an SEM. A film of acetylcellulose was applied to a concrete surface polished to 1 μm. The film was then removed, and double coated with carbon and chromium; the metallized replica was then viewed in an SEM. This technique might also be used to observe microcracking in a process zone.

Bascoul and his colleagues (1989a;1989b) also used a replica technique employing acetylcellulose. Their work showed microcracks in the vicinity of the main crack, but these did not appear to constitute a well-defined process zone. It was also interesting to note (Bascoul et al., 1989b) that microcracks which were visible under load, closed again after failure because of the stress relaxation. This suggests that studies on fracture surfaces after a specimen has broken, may not be able to reveal the presence of a process zone, even if one did exist during crack extension. In subsequent work, Bascoul et al.

(1989a) concluded from observations on the surface of the specimen that "it is difficult to identify a process zone as such. Nothing allows ... [us]... to think that there is a damaged zone ahead of the macrocrack except at the beginning of crack branching." However, in the interior of the specimens, the crack was always preceded by discontinuous microcracks.

6 Electric Resistance Strain Gauge Techniques

Chhuy et al. (1986) used three-dimensional strain gauges to study the damage zone in the vicinity of the crack tip in very large plates. In conjunction with other techniques discussed below, they found a damage zone about 10 mm wide and 100 mm long. Reinhardt and Hordijk (1988) concluded that multiple arrays of strain gauges were suitable for measurement of strain distribution only before the peak stress occurs. However, they were tentatively able to measure a process zone width of 14 mm, using a maximum aggregate size of 2 mm.

John and Shah (1986) used a special brittle Krak-gage to monitor cracking. However, they found that the Krak-gage response was valid only for a single continuous crack; it was invalid in regions of discontinuous crack growth (i.e., in the damage zone).

7 Photoelastic Methods

Lierse and Ringkamp (1983) used a photoelastic technique to study cracking in concrete. A photoelastic foil was glued on to the specimen, and the reflection method was used. Crack widths down to 0.01 mm would be detected, but the results were difficult to interpret.

More recently, however, Van Mier and Nooru-Mohamed (1988) and Van Mier (1988), used a combination of surface deformation measurements and a photoelastic coating technique to study fracture evolution in double-edge notched specimens in direct tension, and in combined tension and shear. They were able to observe that there is a two-stage fracture mechanism:
i)   Perimeter cracking occurs as non-connecting crack branches develop from all sides of the specimen and penetrate to the specimen core.
ii) Subsequently, bending of the intact ligaments between the perimeter cracks occur, with final fracture being due to flexural failure of these ligaments.
They concluded that reflection photoelasticity could indeed be used to study crack localization, crack branching, and crack opening and closing. However, even these results were not able to define a fracture process zone; they could also be interpreted as simply showing the

extension of a macrocrack. On the other hand, Stys (1989) found a process zone about 5 mm in extent, for beams 200-300 mm deep.

8 Mercury Penetration Measurements

Schneider and Diederichs (1983) used mercury penetration methods to detect cracks in the 1 μm to 60 μm range. However, this method can only measure crack widths and crack volumes; it is not sensitive enough to define the process zone.

9 Dye Penetrants

Swartz and Go (1984) and Swartz and Refai (1987) used dye penetration to reveal the position of the crack front after notched beams were broken. A common cleaning agent with a low viscosity was used as the dye. Though this technique can provide an estimate of the shape of the crack front, it is again not sensitive enough to delineate the process zone itself. Bascoul et al. (1987) submerged a specimen of microconcrete in a tank filled with coloured resin diluted with styrene, to which a vacuum was applied. They obtained similar results to these of Swartz and Go (1984) and Swartz and Refai (1987).

10 Infrared Vibrothermography

This novel technique for revealing damage in concrete was described by Luong (1986). Infrared vibrothermography monitors the mechanical response of concrete under compressive load subjected to vibratory excitation. This excitation generates heat when there is energy dissipation by the material, which occurs when the material is excited to beyond its stable reversible limit. This technique permits the observation of progressive damage in a specimen.

11 Ultrasonic Pulse Velocity

Ultrasonic pulse velocity measurements have long been used to try to assess the relative strength and possible large-scale damage in concrete. More recently, however, it has been used to try to assess the size of the damage zone in concrete. Extensive work of this nature has been carried out by Alexander (1985,1988) and Alexander and Blight (1986,1987). They found from their pulse velocity measurements that during fracture of notched concrete beams, the main crack did not propagate until an extensive microcracked zone had developed ahead of it, and reached a

maximum size. Using the definition of stress zones shown
in Fig. 2 (Alexander and Blight, 1986), they found that

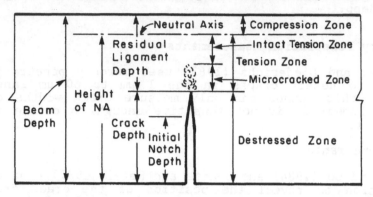

Fig. 2. Definition of stress zones in a concrete beam
         during fracture, after Alexander and Blight
         (1986).

the damage zones were about 40-47% of the residual beam
depth, and this was also reported by Alexander et al.
(1989). Moreover, they concluded that the upper and lower
limits for the microcracked zone were 0.684 and 0.368,
respectively, times the residual ligament depth. These
results are shown in Fig. 3 (Alexander, 1985).
    Chhuy et al. (1986) also used ultrasonic pulse velocity
measurements as part of a programme to study the damage
zone in the vicinity of a crack tip in very large, precracked plates. They found the damage zone to be about
100 mm long and about 10 mm (the maximum aggregate size)
wide. The sonic testing itself seemed to give the length
of the damage zone to about ±20 mm. The sensitivity was
increased markedly when a frequency spectrum analysis was
performed.
    Berthaud (1987,1988) found that both surface damage
and volumetric damage could be assessed using ultrasonic
pulse velocities. However, the degree of accuracy of
these results was uncertain, and it was difficult to
deduce the size of the process zone. More recently,
Reinhardt and Hordijk (1988) concluded that ultrasonic
pulse measurements do not lead to a geometrical description of the damage zone.

12 Demec Gauges

In some of the pulse velocity studies referred to above
(Alexander, 1985; Alexander and Blight, 1986) strain
measurements were made simultaneously, using Demec gauges.
It was found that the use of these gauges provided very

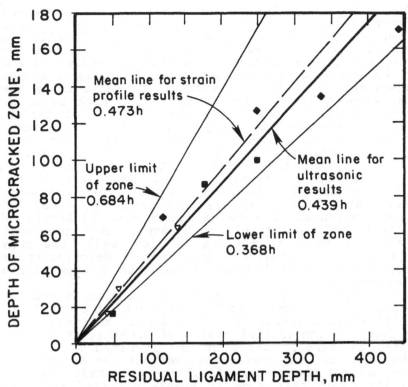

Fig. 3. Dependence of microcracked zone depth on residual ligament depth, from ultrasonic measurements, after Alexander (1985).

similar results to those found using the pulse velocity technique.

13 Acoustic Emission

There has been an increasing use of acoustic emission (AE) measurements to monitor the formation and propagation of cracks in concrete; the work up to 1983 has already been extensively reviewed by Diederichs et al. (1983) and a general review of acoustic emission phenomena in concrete has been prepared by Mindess (1990). In terms of studies specifically to measure the extent of the process zone, however, the results appear to depend upon the specimen geometry, the type of instrumentation used, and the method of analysis. Since this is one of the more promising methods of fracture process zone detection, a brief account of the technique itself is warranted.

Acoustic emissions occur when transient elastic waves are generated by the rapid release of energy from localized sources in the concrete, due primarily to cracking processes. These waves propagate through the concrete, and can be detected as ultrasonic pulses on the surface by piezoelectric transducers. By using a number of transducers to monitor AE events, and determining the time differences between the detection of each event at different transducer positions, the location of the AE event may be determined by using triangulation techniques. Work by Maji and Shah (1990) has indicated that this technique may be accurate to within about 5 mm. Because of signal attenuation, the maximum distance between transducers, or between the transducers and the sources of the AE event, should not be very large. For instance, Berthelot and Robert (1987) required an array of transducers arranged in a 40 cm square mesh to locate AE events reasonably accurately. However, they concluded that "acoustic emission testing is practically the only technique which can provide a quantitative description of the progression in real time of concrete damage within test specimens".

A very extensive series of tests to study the formation and propagation of cracks has been carried out by Maji and Shah (1986,1988a,1988b,1990). They found (Maji and Shah, 1988a) that, in the initial stages of loading, AE signals originated at various parts of the specimen. However, near the peak load and thereafter, most of the AE activity came from around the crack. In their work, Maji and Shah (1986,1990) found that some AE events occurred ahead of the crack tip, while others continued to occur behind the crack tip indicating ligament connections. A process zone extending perhaps 25 mm ahead of the optically visible crack and a larger distance behind the crack tip was deduced from the AE signals.

More sophisticated signal processing techniques (deconvolution techniques) were subsequently used by Maji and Shah (1990) and Maji (1989) to determine the volume, orientation and type of microcrack. The relative amplitudes of AE signals at different transducers were used to distinguish between tension and shear microcracks in mortar and at aggregate-matrix interfaces. Such techniques can be used to provide a detailed picture of the fracture processes occurring in concrete.

On the other hand, Izumi et al. (1986) deduced a process zone size of about 105 mm, in terms of the fictitious crack model. In the work referred to above, Chhuy et al. (1986) found a damage zone about 100 mm long when using AE measurements in conjunction with directional strain gauges and sonic testing. However, in earlier AE work, Chhuy et al. (1979) estimated a damage zone which reached lengths of 500 mm ahead of the crack tip.

Berthelot and Robert (1985a,1985b) have also used AE to study the evolution of damage in concrete. Their test technique involved the use of four AE sensors. With this technique, AE could be used to monitor damage in real time, though there were some limitations on the precision with which the damage zone could be located. They found that a microcracked zone ahead of the continuous crack could be detected as shown schematically in Fig. 4

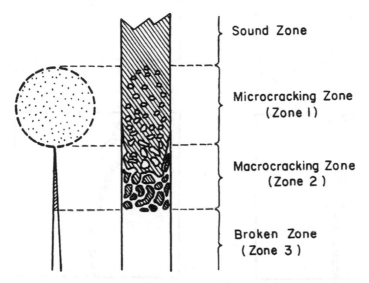

Fig. 4. Damage evolution along the crack, after Berthelot and Robert (1985b).

(Berthelot and Robert, 1985b). This damage zone appeared to grow in size during crack extension, as shown in Fig. 5. Damage zones as large as 120 mm wide and 160 mm long could be detected in this way. Similar results were obtained by Bensouda (1989). Damage evolution of concrete specimens by acoustic emission was also related to failure analysis in a recent paper by Berthelot and Robert (1990).

Recently, an extensive series of studies by Rossi (1986) and Rossi et al. (1989), using five transducers (four placed at the corners of a 40 cm square, and one at the centre of the array), found that the transition from the uncracked to the microcracked zone in concretes of various types is not very clear. However, they determined the maximum width of the microcracked zone to be about 100 mm. They were also able to distinguish between cracks at the cement-aggregate interface (low frequency AE spectra) and cracks in the matrix (high frequency AE spectra).

On the other hand, in the comparative study already referred to, Reinhardt and Hordijk (1988) indicated that

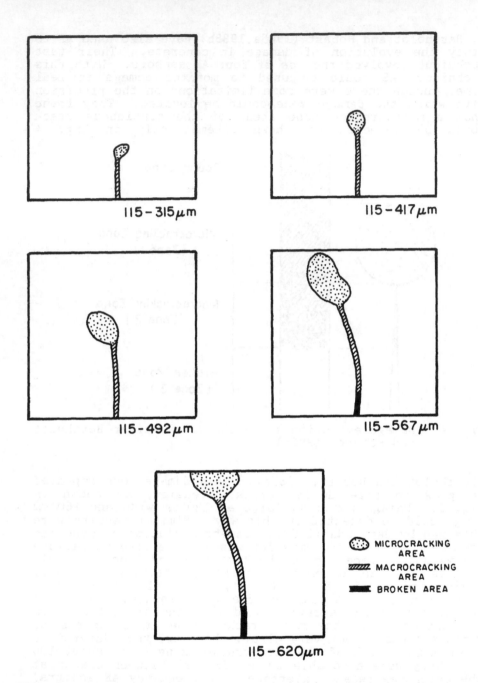

Fig. 5. Damage evolution during the test, in terms of the crack opening displacement, after Berthelot and Robert (1985b).

acoustic emission provides more qualitative than quantitative information about the cracking process. Indeed, they concluded that none of the three techniques they compared (strain and displacement measurements, ultrasonic pulse velocity, and acoustic emission) could "produce full and accurate data on the degree of local damage and the size of the damage zone".

14 Interferometry Techniques

Perhaps the most exciting work on the identification of the process zone is being carried out using interferometry techniques: holographic interferometry, and speckle interferometry. These techniques are lumped together in this section, since they are often performed at the same time. The theory and use of holographic and speckle interferometry techniques has been reviewed extensively by Jacquot and Rastogi (1983), Jacquot (1984), and Iori (1988). While the techniques are difficult to set up experimentally, they are perhaps the most sensitive techniques for the determination of process zone size.

Briefly, laser interferometry techniques can be used to measure the displacement field on the surface of a specimen. Laser Moiré interferometry refers to the analysis of the Moiré patterns obtained by the interference of two grids. One grid is attached to the surface of the specimen and deforms with the specimen as it is loaded; the second (undeformed) grid which acts as a reference grid is obtained by illuminating the surface of the specimen with two laser beams. The resulting interference pattern is recorded on a photographic emulsion. These interference fringes can be analysed to provide a quantitative description of the surface deformations. The theoretical sensitivity of this method is about 0.5 µm, though for materials as rough as concrete, the sensitivity is more realistically in the order of several microns (Jacquot, 1984). The experimental techniques for this method are described in detail in the following chapter.

Holographic interferometry and speckle interferometry are based on the superposition of two holograms or speckle photographs, respectively. The resulting interference patterns represent the difference between the two exposures. Since there is no need to attach a grid on the surface of the specimen and only differences in position are recorded, the sensitivity of these methods is independent of the roughness of the material. With holographic interferometry the sensitivity is about 0.5 µm and with speckle interferometry a sensitivity of 1 to 5 µm can be achieved.

Jacquot and Rastogi (1987) used holographic interferometry sensitive to the in-plane displacements to detect displacements of 0.5 µm; they argued that this technique could be used to detect the damage zone. Maji and Shah (1988a) also found that holographic interferometry could be used to find discontinuities in the displacement field (i.e. cracks) with a sensitivity of about 0.63 μm with speckle interferometry, they were able to achieve a sensitivity of about 1 μm.

Maji (1989) has used holographic interferometry to study mixed mode crack initiation and propagation under compression. He was able simultaneously to measure the opening and sliding displacements along a crack. The holography technique has thus proven to be useful for quantitative measurement of inplane displacements.

Iori et al. (1982) also found that laser Moiré interferometry techniques could be used to measure strains and displacements in concrete, with a sensitivity of about 1 μm. Later tests carried out by Cedolin et al. (1987) using this technique permitted the determination of damage zones perhaps 20-40 mm long ahead of the crack. The zone of diffused microcracking could be detected, as shown schematically in Fig. 6 (after Iori et al., 1982). An

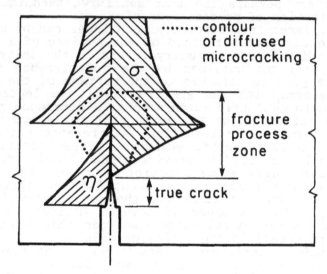

Fig. 6. Fracture process zone as determined using holographic interferometry, after Iori et al. (1982).

example of the quantitative measurements that can be obtained using this technique is given in Fig. 7 (Cedolin et al., 1987). The upper part, (a)-(d), represents an early stage of crack propagation, while the lower part, (e)-(h), represents the stage just before failure.

Dei Poli and Iori (1986) used laser Moiré interferometry to obtain lines of isodeformation at different stages of loading, as shown in Fig. 8. On the assumption

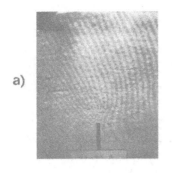

a)

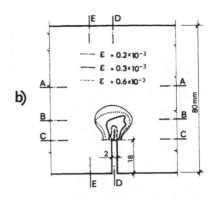

b)

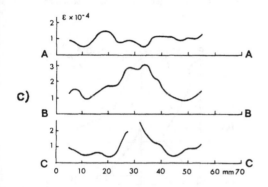

c)

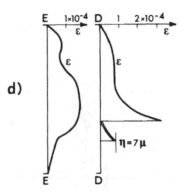

d)

Fig. 7.   Results obtained using optical interferometry
with laser light, after Cedolin et al. (1987):
(a),(e) fringe configurations; (b),(f) equal
strain contours and developing crack; (c),(g)
strain distribution along longitudinal sections;
(d),(h) strain distribution along transverse
cross sections and crack opening profile.

Note: Figs. (a)-(d) refer to an intermediate load
level, while Figs. (e)-(h) refer to a load level
close to failure.

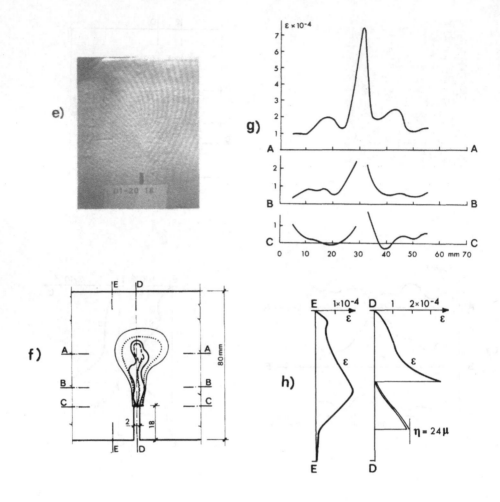

Fig. 7. (Continued)

that the process zone is bounded by the line of isodeformation representing a strain of 0.15%, the evolution of this zone as loading progresses can be followed. From Fig. 8, it can be seen that the amplitude of the process zone is about 50 mm.

On the other hand, holographic interferometry studies by Ferrara and Morabito (1989) with specimens in four-point shear showed that while microcracks could be detected ahead of the macrocrack, it was not clear that these microcracks denoted a well-defined process zone.

Castro-Montero et al. (1990) developed a multiple sensitivity vector holographic setup to measure both crack

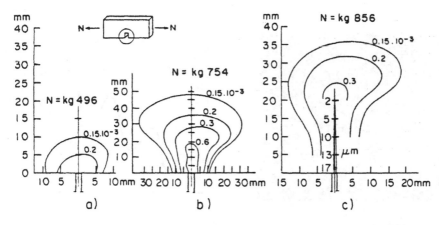

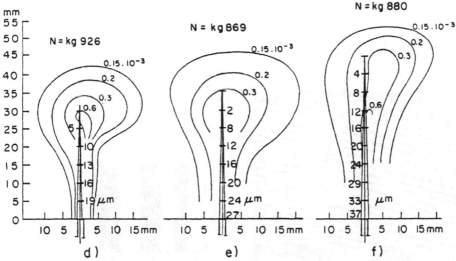

Fig. 8. Isodeformation contours at various stages of
loading, after Dei Poli and Iori (1986).

opening displacements and strain fields around the crack
trajectories. Image analysis techniques were used for
isolation of the interferometric effect from the sandwich
holograms (Fig. 9). In Fig. 9, image 1 corresponds to the
sandwich interferogram and image 2 corresponds to the
average intensity of the two holograms acquired separately. Further image processing was used for accurate and
consistent fringe counts.

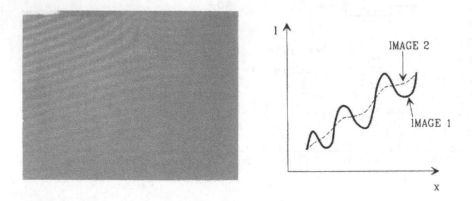

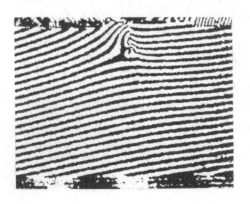

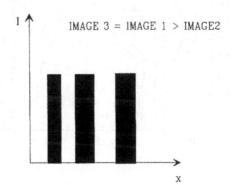

Fig. 9.   Image analysis techniques used for isolation of
the interferometric effect from a sandwich hologram, after Castro-Montero et al. (1990). Image 1
corresponds to the sandwich interferogram; Image
2 corresponds to the average intensity of the two
holograms acquired separately.

The existence of experimentally observed tensile strains behind the crack tip was associated with the presence of tensile forces transmitted through the crack. A definition of the fracture process zone was proposed based on the difference between experimentally measured and LEFM strain fields. Deviations from the linear elastic solution show a relatively small zone of constant size of nonlinearity in front of the crack tip (zones A in Fig. 10). Zones B in Fig. 10 can be defined as the wake of the fracture process zone which encloses the area where extensive microcracking has occurred and tensile forces are still transmitted through the crack.

Regnault and Brühwiler (1988) also used a real time holographic Moiré technique to study the growth of the fracture process zone. They concluded that the process zone consists of two parts: an elliptical region within which the material behaves like a continuum, and a narrow band of disrupted material in which load is transferred due to aggregate interlock. A significant process zone existed prior to the peak load; at the peak load, the length of the process zone was about half the length of the tensile zone. However, it was not possible to locate, or even to define, the "tip" of a tension-free crack.

Hansen (1989) used TV-holography for real-time observations of microcracking; he was able to observe microcracks down to a width of about 0.1 μm. However, he was unable to observe any zone of extensive microcracking, due to limitations in the resolution of the system.

Xu and Zhao (1989) used laser speckle photography to examine the notched beams in bending. They observed that fracture process zones occurred as long, narrow bands, with maximum widths of about 1/2 the maximum aggregate size, and lengths up to 25 mm (for a specimen depth of 100 mm). Ansari (1986) used speckle interferometry to study the extent of the microcracked zone and the associated displacements. He found a microcracked zone extending about 30 mm ahead of the visible crack. Interestingly, he noted that displacements about 17-26 mm ahead of the crack tip were greater than those right at the crack tip, suggesting that the growth of microcracks in the process zone may be governed primarily by stress concentrations due to large aggregate particles, and this may be largely independent of what is going on right at the crack tip.

White light Moiré interferometry was used by Du et al. (1987) to locate the fracture process zone ahead of the crack tip. The use of a motor driven camera at five frames/ second to record the different Moiré patterns gave an indication of the development of the process zone as the crack extended. The technique was apparently very sensitive, with the frequency of the reference grating being 1200 lines/mm. In subsequent work, Du et al. (1989) reported a process zone length of up to 100 mm, for both DCB and beam specimens. Similarly, Raiss et al. (1989)

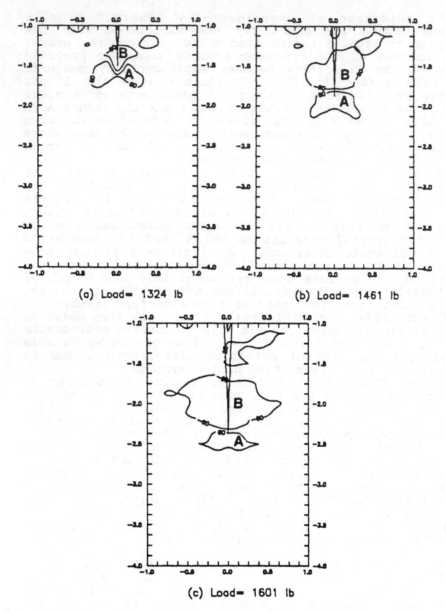

(a) Load= 1324 lb

(b) Load= 1461 lb

(c) Load= 1601 lb

Fig. 10. The fracture process zone, as determined from the difference between experimentally measured and LEFM strain fields, after Castro-Montero et al. (1990). Zones A show a small zone of non-linearity in front of the crack tip; Zone Bs are defined as the wake of the fracture process zone.

carried out uniaxial tests in a very stiff closed-loop machine. Using Moiré interferometry they observed that fracture occurred by the initiation and gradual propagation of the process zone. They observed these zones to be about 10 mm wide, irregularly shaped, and oriented approximately perpendicular to the tensile stress. Lengths ranged up to 20-30 mm.

15 Multi-Cutting Techniques

A more direct determination of the fracture process zone in concrete can be achieved with a multi-cutting technique. In this method, thin strips of the specimen are cut normal to the crack path, in the vicinity of the crack tip, and the bending stiffness of each strip is then measured as a function of the distance away from the tip. Since microcracking will reduce the stiffness, this allows an estimate of the fracture process zone size to be made. Foot et al. (1987) found a process zone of about 30 to 40 mm for a compact tension geometry in fibre reinforced cements.

A variation of this technique was developed by Hu and Wittmann (1989a,1990). In their technique, the bridging stress transferred within the fracture process zone is removed step by step by cutting with a diamond saw along the plane of the original cast (or pre-cut) crack through the microcracked region, and determining the compliance of the specimen at each step. Since the change in compliance is due primarily to the removal of "bridging" stresses behind the crack tip, the size of the process zone can be estimated. Hu and Wittmann (1989b) also developed an analytical method to be used with the multi-cutting technique, to determine the bridging stresses transferred within the process zone. Their experimental results indicated that, for WOL specimens with a ligament length of 110 mm, the length of the process zone in mortar varied from about 43 mm to 12 mm as the crack approached the boundary. They concluded that "the length of the fracture process zone is not a material constant but depends on the actual stress gradient due to the limited specimen geometry" (Hu and Wittmann, 1989a).

16 Numerical Methods

The discussion above has dealt entirely with experimental techniques. However, Schorn and Rode (1987) have suggested a numerical simulation method to model the process zone. Their method uses a large number of struts in an equidistant lattice point arrangement as a three-dimensional mechanical model. It was suggested that this technique could simulate microcrack growth up to the point of total collapse of the system.

Alvaredo *et al*. (1989) simulated the fictitious crack model using a finite element formulation. This could be used to predict the local material response, and hence the presence of a process zone, but the method was very sensitive to the exact assumptions made. Gopalaratnam and Ye (1989) also used a numerical scheme based on the fictitious crack model to study the process zone. They calculated that the process zone reaches a steady state length which is "somewhat" independent of specimen size, but which is geometry dependent. A process zone length of about 40-80 mm was predicted.

17 Conclusions

As may be seen from the above review, more than a dozen techniques have been used seriously to try to determine the extent of the process zone in concrete. In the absence of any standardized test geometry, it is not surprising that estimations of the size of the process zone range from only a few mm to perhaps 500 mm. The most sensitive techniques appear to be those based on interferometry, using laser light, but unfortunately these are also the most difficult to set up experimentally, and probably the most expensive. For reasonably large specimens, both acoustic emission and ultrasonic pulse velocity techniques would appear to give sensible results. For the time being, it would seem that further work must be carried out, in a systematic way, with a number of the techniques described, in order for us to achieve a reasonable understanding of the development and extent of the process zone. Given the variability in results, however, it is difficult to escape the conclusion that the fracture process zone is not a fundamental material property (at least for laboratory-size specimens), but depends on the specimen geometry and method of loading.

18 Acknowledgements

The author would like to acknowledge with thanks the many comments that he received from other members of RILEM Committee 89-FMT on earlier drafts of this chapter and some manuscripts still in draft form. In particular, he would like to thank Dr. J. Kasperkiewicz, Dr. Z.P. Bazant, Dr. M.G. Alexander, Dr. L. Cedolin, Dr. A.K. Maji, Dr. S.E. Swartz, Dr. B.L. Karihaloo, Dr. J.M. Berthelot, Dr. P. Rossi, Dr. S.P. Shah and Dr. A. Castro-Montero for their extensive contributions.

This work was supported in part by a grant from the Natural Sciences and Engineering Research Council of Canada.

19 References

Alexander, M.G. (1985) Fracture of Plain Concrete - A
Comparative Study of Notched Beams of Varying Depth,
Ph.D. **Thesis**, University of the Witwatersrand,
Johannesburg, South Africa.

Alexander, M.G. (1988) Use of Ultrasonic Pulse Velocity
for Fracture Testing of Cemented Materials, **Journal
of Cement, Concrete and Aggregates**, ASTM, Vol. 10,
No. 1, Summer, pp. 9-14.

Alexander, M.G. and Blight, G.E. (1986) The Use of Small
and Large Beams for Evaluating Concrete Fracture
Characteristics, in F.H. Wittmann, (ed.), **Fracture
Toughness and Fracture Energy of Concrete**, Elsevier
Science Publishers B.V., Amsterdam, pp. 323-332.

Alexander, M.G. and Blight, G.E. (1987) Characterization
of Fracture in Cemented Materials, **private communication**.

Alexander, M.G., Tait, R.B. and Gill, L.M. (1989)
Characterization of Microcracking and Crack Growth in
Notched Concrete and Mortar Beams Using the J-Integral
Approach, in S.P. Shah, S.E. Swartz and B. Barr (eds.),
Fracture of Concrete and Rock: Recent Developments,
Elsevier Applied Science, London, pp. 317-326.

Alvaredo, A.M., Hu, X.Z. and Wittmann, F.H. (1989) A
Numerical Study of the Fracture Process Zone, in S.P.
Shah, S.E. Swartz and B. Barr (eds.), **Fracture of
Concrete and Rock: Recent Developments**, Elsevier
Applied Science, London, pp. 51-60.

Ansari, F. (1986) Analysis of Micro-Cracked Zone in
Concrete, in F.H. Wittmann, (ed.), **Fracture Toughness
and Fracture Energy of Concrete**, Elsevier Science
Pulishers B.V., Amsterday, pp. 229-240.

Baldie, K.D. and Pratt, P.L. (1986) Crack Growth in
Hardened Cement Paste, in S. Mindess and S.P. Shah,
(eds.), **Cement-Based Composites: Strain Rate Effects
on Fracture**, Materials Research Society Symposia
Proceedings Vol. 64, Materials Research Society,
Pittsburgh, pp. 47-61.

Bascoul, A., Detriche, C.H., Ollivier, J.P. and
Turatsinze, A. (1989a) Microscopical Observation of the
Cracking Propagation in Fracture Mechanics for Con-
crete, in S.P. Shah, S.E. Swartz and B. Barr (eds.),
Fracture of Concrete and Rock: Recent Developments,
Elsevier Applied Science, London, pp. 327-336.

Bascoul, A., Kharchi, F. and Maso, J.C. (1987) Concerning
the Measurement of the Fracture Energy of a
Micro-Concrete According to the Crack Growth in a Three
Point Bending Test on Notched Beams, in S.P. Shah and
S.E. Swartz (eds.), **SEM/RILEM International
Conference on Fracture of Concrete and Rock**, Houston,
Society of Experimental Mechanics, Bethel CT, pp.
631-643.

Bascoul, A., Ollivier, J.P. and Poushanchi, M. (1989b) Stable Microcracking of Concrete Subjected to Tensile Strain Gradient, **Cement and Concrete Research**, Vol. 19, No. 1, pp. 81-88.

Bazant, Z.P. (1988) Letter to the author, October 26.

Bazant, Z.P. and Oh, B.H. (1983) Crack Band Theory for Fracture of Concrete, **Materiaux et Constructions**, <u>16</u> (93) 155-177.

Bensouda, M. (1989) Contribution a L'Analyse par Emission Acoustique de L'Endommagement dans le Beton, Ph.D. **Thesis**, Université du Maine, France.

Berthaud, Y. (1987) An Ultrasonic Testing Method: An Aid for Material Characterization, in Proceedings, **International Conference on the Fracture of Rock and Concrete**, Houston, pp. 644-654.

Berthaud, Y. (1988) Mesure de L'Endommagement du Beton Par Une Methode Ultrasonore, Ph.D. **Thesis**, Univerisité de Paris 6, France.

Berthelot, J.-M. and Robert, J.-L. (1985a) Damage Process Characterization in Concrete by Acoustic Emission, presented at the **Second International Conference on Acoustic Emission**, Lake Tahoe, Nevada, October.

Berthelot, J.-M. and Robert, J.-L. (1985b) Application de l'Emission Acoustique aux Mecanismes d'Endommagement du Beton, **Bulletin de Liaison des Ponts et Chaussees**, <u>140</u>, Nov.-Dec., 101-111.

Berthelot, J.-M. and Robert, J.-L. (1987) Modelling Concrete Damage by Acoustic Emission, **Journal of Acoustic Emission**, <u>6</u> (1) 43-60.

Berthelot, J.-M. and Robert, J.-L. (1990) Damage Evaluation of Concrete Test Specimens Related to Failure Analysis, ASCE, **Journal of Engineering Mechanics**, <u>116</u> (3) 587-604.

Castro-Montero, A., Shah, S.P. and Miller, R.A. (1990) Measurement of Strain Fields in Fracture Process Zone of Mortar, ASCE, **Journal of Engineering Mechanics** (in press).

Cedolin, L., Dei Poli, S. and Iori, I. (1987) Tensile Behaviour of Concrete, **Journal of Engineering Mechanics**, <u>113</u> (3) 431-449.

Chhuy Sok, Baron, J. and Francois, D. (1979) Mecanique de la Rupture Applique au Beton Hydraulique, **Cement and Concrete Research**, <u>9</u> (5) 641-648.

Chhuy, S., Cannard, G., Robert, J.L. and Acker, P. (1986) Experimental Investigations into the Damage of Cement Concrete with Natural Aggregates, in A.M. Brandt and I.H. Marshall (eds.), **Brittle Matrix Composites 1**, Elsevier Applied Science, London, pp. 341-354.

Dei Poli, S. and Iori, I. (1986) Osservazioni e Rilievi sul Comportamento a Trazione dei Calcestruzzi: Analisi di Risultanze Sperimentali, **Studi e Ricerche**, Politecnico di Milano, <u>8</u> 35-62.

Diamond, S. and Bentur, A. (1985) On the Cracking in Concrete and Fibre-Reinforced Cements, in S.P. Shah, (ed.), **Applications of Fracture Mechanics to Cementitious Composites**, Martinus Nijhoff Publishers, Dordrecht, pp. 87-140.

Diamond, S., Mindess, S. and Lovell, J. (1983) Use of a Robinson Backscatter Detector and 'Wet Cell' for Examination of Wet Cement Paste and Moprtar Specimens Under Load, **Cement and Concrete Research**, 13 (1) 107-113.

Diederichs, U., Schneider, U. and Terrien, M. (1983) Formation and Propagation of Cracks and Acoustic Emission, in F.H. Wittmann, (ed.), **Fracture Mechanics of Concrete**, Elsevier Science Publishers B.V., Amsterdam, pp. 157-205.

Du, J., Hawkins, N.M. and Kobayashi, A.S. (1989) A Hybrid Analysis of Fracture Process Zone in Concrete, in Shah, S.P., Swartz, S.E. and Barr, B. (eds.), **Fracture of Concrete and Rock: Recent Developments**, Elsevier Applied Science, London, pp. 297-306.

Du, J.J., Kobayashi, A.S. and Hawkins, N.M. (1987) Fracture Process Zone of a Concrete Fracture Specimen, in Proceedings, **International Conference on Fracture Concrete and Rock**, Houston, pp. 280-286.

Eden, N.B. and Bailey, J.E. (1986) Crack Tip Processes and Fracture Mechanism in Hardened Hydraulic Cements, in Proceedings, **8th International Congress on the Chemistry of Cement**, Rio de Janeiro, Vol. III, pp. 382-388.

Ferrara, G. and Morabito, P. (1989) A Contribution of the Holographic Interferometry to Studies on Concrete Fracture, in S.P. Sha, S.E. Swartz and B. Barr (eds.), **Fracture of Concrete and Rock: Recent Developments**, Elsevier Applied Science, London, pp. 337-346.

Foote, R., Mai, Y. and Cotterell, B. (1987) Process Zone Size and Crack Growth Measurements in Fiber Cements, in S.P. Shah and G.B. Batson (eds.) **Fiber Reinforced Concrete Properties and Applications**, SP-105, American Concrete Institute, Detroit, pp. 55-70.

Gopalaratnam, V.S. and Ye, B.S. (1989) Numerical Studies of the Fracture Process Zone in Concrete, in S.P. Shah, S.E. Swartz and B. Barr (eds.), **Fracture of Concrete and Rock: Recent Developments**, Elsevier Applied Science, London, pp. 81-90.

Hansen, E.A. (1989) A Holographic Real Time Study of Crack Propagation in Concrete, **Cement and Concrete Research**, Vol. 19, pp. 611-620.

Hu, X.Z. and Wittmann, F.H. (1989a) Fracture Process Zone and K_r-Curve of Hardened Cement Paste and Mortar, in S.P. Shah, S.E. Swartz and B. Barr (eds.), **Fracture of Concrete and Rock: Recent Developments**, Elsevier Applied Science, London, pp. 307-316.

Hu, X.A. and Wittmann, F.H. (1989b) An Analytical Method to Determine the Bridging Stress Transferred within the Fracture Process Zone, private communication.

Hu, X.Z. and Wittmann, F.H. (1990) An Experimental Method to Determine the Extension of the Fracture Process Zone, submitted to the ASCE Journal of Materials in Civil Engineering.

Iori, I. (1988) Process Zone in Normal Concrete, private communication.

Iori, I., Lu, H., Marozzi, C.A. and Pizzinato, E. (1982) Metodo per la Determinazione dei Campi dei Spostamento nei Materiali Eterogenei (Conglomerati Naturali ed Artificiali) a Bassa Resistenza Specifica a Trazione, L'Industria Italiano Del Cemento, No. 4, 275-280.

Izumi, M. Mihashi, H. and Nomura, N. (1986) Acoustic Emission Technique to Evaluate Fracture Mechanics Parameters of Concrete, in F.H. Wittmann, (ed.), Fracture Toughness and Fracture Energy of Concrete, Elsevier Science Publishers B.V., Amsterdam, pp. 259-268.

Jacquot, P. (1984) Interferometry in Scattered Coherent Light Applied to the Analysis of Cracking in Concrete, in A. Carpinteri and A.R. Ingraffea, (eds.), Fracture Mechanics of Concrete: Material Characterization and Testing, Martinus Nijhoff Publishers, The Hague, pp. 161-194.

Jacquot, P. and Rastogi, P.K. (1983) Speckle Metrology and Holographic Interferometry Applied to the Study of Cracks in Concrete, in F.H. Wittmann, (ed.), Fracture Mechanics of Concrete, Elsevier Science Publishers B.V., Amsterdam, pp. 113-155.

Jacquot, P. and Rastogi, P.K. (1987) Private communication.

John, R. and Shah, S.P. (1986) Fracture of Concrete Subjected to Impact Loading, Cement, Concrete, and Aggregates, 8 (1) 24-32.

Karihaloo, B.L. and Nallathambi, P. (1989) An Improved Effective Crack Model for the Determination of Fracture Toughness of Concrete, Cement and Concrete Research, 19 (4) 603-610.

Karihaloo, B.L. and Nallathambi, P. (1990) Size Effect Prediction from Effective Crack Model for Plain Concrete, in preparation.

Kasperkiewicz, J. (1988) Letter to the author, November 22.

Knab, L.I., Jennings, H., Walker, J.N., Clifton, J.R. and Grimes, J.W. (1986) Techniques to Observe the Fracture Zone in Mortar and Concrete, in F.H. Wittmann, (ed.), Fracture Toughness and Fracture Energy of Concrete, Elsevier Science Publishers B.V., Amsterdam, pp. 241-247.

Knab, L.I., Walker, J.N., Clifton, J.R. and Fuller, E.R., Jr. (1984) Fluorescent Thin Sections to Observe the Fracture Zone in Mortar, Cement and Concrete Research, 14 (3) 339-344.

Kobayashi, A.S., Hawkins, N.M., Barker, D.B. and Liaw, B.M. (1985) Fracture Process Zone of Concrete, in S.P. Shah (ed.) Applications of Fracture Mechanics to Cementitious Composites, Martinus Nijhoff Publishers, Dordrecht, pp. 25-50.

Lierse, J. and Ringkamp, M. (1983) Investigations into Cracked Reinforced Concrete Structural Elements with the Aid of Photoelastic Methods, in F.H. Wittmann, (ed.), Fracture Mechanics of Concrete, Elsevier Applied Science Publishers B.V., Amsterdam, pp. 95-111.

Luong, M.P. (1986) Infrared Vibrothermography of Plain Concrete, in F.H. Wittmann, (ed.), Fracture Toughness and Fracture Energy of Concrete, Elsevier Applied Science Publishers B.V., Amsterdam, pp. 249-257.

Maji, A.K. (1989) Study of Concrete Fracture Using Acoustic Emission and Laser Holography, Ph.D. Thesis, Northwestern University, Evanston, IL, June.

Maji, A.K. and Shah, S.P. (1986) A Study of Fracture Process of Concrete Using Acoustic Emission, Proc., Soc. Expt. Mech. Spring Conference, New Orleans, June 8-13.

Maji, A.K. and Shah, S.P. (1988a) Application of Acoustic Emission and Laser Holography to Study Microfracture in Concrete, ACI-SP-112 Nondestructive Testing, American Institute, pp. 83-109.

Maji, A. and Shah, S.P. (1988b) Initiation and Propagation of Bond Cracks as Detected by Laser Holography and Acoustic Emission, in S. Mindess and S.P. Shah (eds.) Bonding in Cementitious Materials, Materials Research Society Symposium Proceedings, Vol. 114, Materials Research Society, Pittsburgh, pp. 55-64.

Maji, A.K. and Shah, S.P. (1990) Process Zone and Acoustic Emission Measurements in Concrete, to be published in Experimental Mechanics (SEM Paper No. 3609).

Mindess, S. (1988) The Fracture Process Zone - Myth or Reality?, in Proceedings of the Engineering Foundation Conference on Advances in Cement Manufacture and Use, Potosi, Missouri, pp. 17-22.

Mindess, S. (1990) Acoustic Emission in V.M. Malhotra and N. Carino (eds.), Handbook of Nondestructive Testing of Concrete, CRC Press, Inc., Boca Raton, Florida (in press).

Mindess, S. and Diamond, S. (1982a) The Cracking and Fracture of Mortar, Materiaux et Constructions, 15 (86) 107-113.

Mindess, S. and Diamond, S. (1982b) A Device for Direct Observation of Cement Paste or Mortar Under Compressive Loading Within a Scanning Electron Microscope, Cement and Concrete Research, 12 (5) 569-576.

Najjar, W.S. and Hover, K.C. (1988) Modification of the X-Radiography Technique to Include a Contrast Agent for Identifying and Studying Microcracking in Concrete, Cement, Concrete and Aggregate, 10 (1) 15-19.

Raiss, M.E., Dougill, J.W. and Newman, J.B. (1989) Observation of the Development of Fracture Process Zones in Concrete, in S.P. Shah, S.E. Swartz and B. Barr, Fracture of Concrete and Rock: Recent Developments, Elsevier Applied Science, London, pp. 243-253.

Regnault, Ph. and Brühwiler, E. (1988) Holographic Interferometry for the Determination of Fracture Process Zone in Concrete, in Proceedings, Int. Conf. on Fracture and Damage of Concrete and Rock, Vienna, (to be published).

Reinhardt, H.W. and Hordijk, D.A. (1988) Various Techniques for the Assessment of the Damage Zone Between Two Saw Cuts, presented at the France-U.S. Workshop, Strain Localization and Size Effect Due to Cracking and Damage, Cachan, France, September, 12 pp.

Ringot, E., Ollivier, J.P. and Maso, J.C. (1987) Characterization of Initial State of Concrete with Regard to Microcracking, Cement and Concrete Research, 17 (3) 411-419.

Rossi, P. (1986) Fissuration due Beton: due Materiaux a la Structure, Application de la Mecanique Lineaire de Rupture, Ph.D. Thesis, Ecole Nationale des Ponts et Chaussees, France.

Rossi, P., Robert, J.L., Gervais, J.P. and Bruhat, D. (1989) Acoustic Emission Applied To Study Crack Propagation in Concrete, Materiaux et Constructions (Paris), Vol. 22, No. 131, pp. 374-383.

Schneider, U. and Diederichs, U. (1983) Detection of Cracks by Mercury Penetration Measurements, in F.H. Wittmann, (ed.), Fracture Mechanics of Concrete, Elsevier Science Publishers B.V., Amsterdam, pp. 207-222.

Schorn, H. and Rode, U. (1987) 3-D Modelling of Process Zone in Concrete by Numerical Simulation, in Proc., International Conference on Fracture of Concrete and Rock, Houston, pp. 308-316.

Shah, S.P. (1988) Fracture Toughness of Cement-Based Materials, Materiaux et Constructions, 21 (122) 145-150.

Slate, F.O. (1983a) X-Ray Technique for Studying Cracks in Concrete, with Emphasis on Methods Developed and Used at Cornell University, in F.H. Wittmann (ed.) Fracture Mechanics of Concrete, Elsevier Science Publishers B.V., Amsterdam, pp. 85-93.

Slate, F.O. (1983b) Microscopic Observation of Cracks in Concrete, with Emphasis on Techniques Developed and Used at Cornell University, in F.H. Wittmann (ed.) **Fracture Mechanics of Concrete**, Elsevier Science Publishers B.V., Amsterdam, pp. 75-83.

Slate, F.O. and Hover, K.C. (1984) Microcracking in Concrete, in A. Carpinteri and A.R. Ingraffea (eds.) **Fracture Mechanics of Concrete: Material Characterization and Testing**, Martinus Nijhoff Publishers, The Hague, pp. 137-159.

Stroeven, P. (1988a) Characterization of Microcracking in Concrete, preprint, **RILEM Conference on Cracking and Durability of Concrete**, Saint Rémy le Chevreuses.

Stroeven, P. (1988b) Some Observations on Microcracking in Concrete Subjected to Various Loading Regimes, Proc., **International Conference on Fracture and Damage of Concrete and Rock**, Vienna, (in press).

Stys, D. (1989) Numerical Analysis of the Stress Field Parameters in the Fracture Process Zone in Concrete, in A.M. Brandt and I.H. Marshall (eds.), **Brittle Matrix Composites 2**, Elsevier Applied Science, London, pp. 134-143.

Swartz, S.E. and Go, C-G. (1984) Validity of Compliance Calibration to Cracked Concrete Beams in Bending, **Experimental Mechanics**, 24 (2) 129-134.

Swartz, S.E. and Refai, T.M.E. (1987) Influence of Size Effects on Opening Mode Fracture Parameters for Precracked Concrete Beams in Bending, in Proceedings, **International Conference on Fracture of Concrete and Rock**, Houston, pp. 403-417.

Tait, R.B. and Garrett, G.G. (1986) In Situ Double Torsion Fracture Studies of Cement Mortar and Cement Paste Inside a Scanning Electron Microscope, Cem. **Concrete Res.**, 16 (2) 143-155.

Thouless, M.D. (1988) Bridging and Damage Zones in Crack Growth, **Journal of the American Ceramic Society**, 71 (6) 408-413.

Van Mier, J.G.M. (1988) Fracture Study of Concrete Specimens Subjected to Combined Tensile and Shear Loading, Proceedings, **International Conference on Measurements and Testing in Civil Engineering**, Lyon-Villeurbanne, 1988 (in press).

Van Mier, J.G.M. and Nooru-Mohamed, M.B. (1988) Geometrical and Structural Aspects of Concrete Fracture, Proceedings, **International Conference on Fracture and Damage of Concrete and Rock**, Vienna, (in press).

Xu, Shilang and Zhao, Guofan (1989) A Study on Fracture Process Zones in Concrete by Means of Laser Speckle Photography, in A.M. Brandt and I.H. Marshall (eds.), **Brittle Matrix Composites 2**, Elsevier Applied Science, London, pp. 333-341.

6 LASER INTERFEROMETRY METHODS

A.K. MAJI
University of New Mexico, Albuquerque, USA
S.P. SHAH
Northwestern University, Evanston, USA

1 Holographic Interferometry (HI)

1.1 Introduction

The concepts of Holography and Laser Interferometry have been around since the pioneering work of Dennis Gabor in the earlier part of the century. Smith (1969) provided an overview of the technique. Hariharan (1984) summarized much of the work in later periods and a good deal of the mathematical background necessary for a thorough understanding of the physical principles. A good deal of practical insight and engineering application was provided by Abramson (1981) and Vest (1979). Jacquot and Rastogi (1983), Jacquot (1984) and Iori (1988) have summarized various laser interferometry techniques and their applications to understanding behaviour of concrete.

In Holographic Interferometry, the first step is the making of a 'Hologram'. This is accomplished with the set-up shown in Figure 1. A laser beam is divided into two parts by a partially reflecting mirror called 'Beam-Splitter'. A part of the beam called the 'Object Beam' is projected on the object after it is expanded by a lens. This light is then diffusely reflected from the object's surface and comes to a photographic plate. This plate has to be highly sensitive (greater than 2000 lines/mm) to be able to record the intensity variations due to interference accurately. Meanwhile, the other beam called the 'Reference Beam' is expanded by another lens and goes directly to the same photographic plate. Since the laser light is monochromatic, the object and the reference lights interfere on the plane of the photographic plate. If the plate is exposed for a while (from a fraction of a second to a few minutes) and developed, it contains information on the interference that is not visible to the naked eye. However, if the photographic plate is repositioned at its exact original position during exposure and the object light is shut off, one can see the virtual image of the object in the exact position of the original object. This image is caused by the diffraction of the reference light by the developed and repositioned holographic plate.

If the object light is also kept on, it is possible to see two objects simultaneously while looking through the plate. One object is the actual object with the object light shining on it. The other is the virtual image or Holographic image of the original object. Since the two objects are exactly superposed, it is not possible to distinguish the two. If, however, the actual object is now deformed or shifted in position, the light from the two objects reaching the eye simultaneously interfere to form a fringe pattern. It is now possible to see the object surface containing a fringe pattern which represents the displacement field on the object surface. These fringes correspond to the component of surface displacements along a sensitivity vector which, for each point on the surface is the bisector of the angle of illumination and the angle of observation of that object (described later in section 1.4).

In the process described above, as the object is stressed, the fringe pattern changes corresponding to displacement of the individual points on the object. The number of fringes increase gradually with increasing displacements until they are too close for the eye

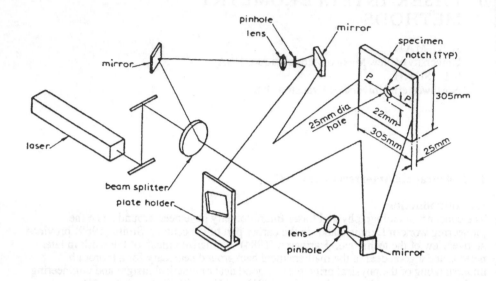

Figure 1. Holographic Interferometry Set-up (Miller et al, 1988)

to distinguish. These fringes are continuous across the object except at a crack where the displacement field has a discontinuity. The cracks are clearly visible as abrupt breaks or kinks in the fringe pattern. Since the fringes are caused by interference of laser light, the sensitivity of measurements by this technique is a fraction of the wavelength of the laser used (0.63 microns for Helium-Neon,'He-Ne' laser).

1.2 Technical details

Additional mirrors are used to adjust directions and distances. A 'Pin-Hole' (typically 5 to 25 microns in diameter) is used to filter out spurious fringes from both the object and reference beams caused by dust particles, etc. The pin-hole is positioned in front of the expanding lens (a microscope objective) so that the beam is focused to pass through the pin-hole before expansion in a process called 'Spatial Filtering'.

The process of looking at a gradually deforming object while simultaneously viewing its holographic image is called 'Real-Time HI'. Alternately, the technique called 'Double Exposure HI' is used. In this technique, the holographic plate is exposed once at the initial and then at a subsequent loading stage. Two holograms are hence made on the same plate with two different loads on the object. When this plate is developed and illuminated with the reference light, the image of the object is formed and it contains a fringe pattern corresponding to the displacement between the two loading stages.

The Real Time technique allows the viewer to study the gradual change in crack and deformation patterns and provides a technique for continuous data acquisition. Since the fringe patterns change as soon as the specimen is loaded again, the data reduction has to be done immediately. Alternately, a video camera could be used continually to monitor the changes in fringe patterns. The Double Exposure method creates a permanent recording of that information between two stages of loading only. This recording is permanently imprinted on the hologram and the fringe patterns can be observed at a later time using some other reference light source. Double Exposure images can be brighter since the entire laser power can be focused on the holographic plate unlike in Real Time technique, where the object has to be well illuminated. A variety of industrial applications in Civil and Mechanical Engineering have been presented by Ebbeni (1982).

The major shortcoming of Holographic Interferometry is the necessity for vibration isolation of the setup. In order for the interference to occur at the holographic plate the optical components (mirrors, lenses etc.) should be vibration free. A general guideline is that, any vibration or motion that causes a shift in the object or the reference beam of a quarter wavelength will destroy the hologram. This motion is considered over the exposure time of the holographic plate. Therefore, short exposure times are better since the chances of spurious motions or vibrations are less. Vibration isolation tables are used to mount all the components and even the loading device. Typically, most structural testing laboratories are located in the basements where vibration problems are minimal compared to upper storeys. Night time experimentation is recommended because of low traffic (pedestrian and automobiles).

In the absence of an isolation table or for a testing machine that is separate from the table, various alternatives have been discussed by Erf (1974). The best but by far more expensive process is to use a pulsed laser (Ruby) where the exposure time is in nanoseconds and any vibration within that time is negligible.

The vibration characteristics of a setup can be studied by setting up a 'Michelson Interferometry' test. In this test, the laser light is divided into two parts by a beam-splitter. The two rays of light are reflected back to the beam-splitter, and form an interference pattern on the screen (Fig. 2) after being expanded by a lens. Any vibration between the two mirror positions can be seen from the vibration or instability in the fringe pattern visible on the screen. The setup is simple and can be applied to test a floor, vibration isolation table or testing machine for a quick and easy estimate of its vibration characteristics.

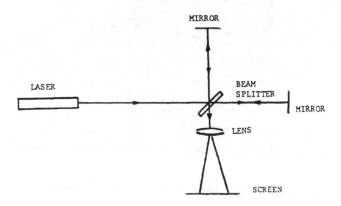

Figure 2.   Set-up for Michelson Interferometry

The second limitation comes from the 'coherence length' of the laser used. This length (typically a few cm. long, for He-Ne lasers), sets limitation on the optical setup and the size of the specimen that can be tested. The difference between the total object light path length and the total reference light path length (from the beam splitter to the plate) should be less than the coherence length. If the object is large, it is difficult to satisfy this condition for all the different points on the object which naturally have different object distances.

1.3 Review of HI applications for crack detection

Maji and Shah (1988) used conventional HI to detect crack initiation and propagation in model concrete with a variety of microstructures. Crack initiation from aggregate interface, voids, and crack propagation in mortar matrix were studied. Tasdemir, Maji and Shah (1990) studied crack initiation at aggregate-mortar interface and subsequent propagation into the mortar matrix. Mixed mode crack initiation and propagation theories were used to understand the experimentally observed phenomena. Aggregate-mortar interfacial friction and traction forces across a propagating crack were found to have significant effect on fracture behaviour of the material.

Stroeven (1989) applied HI to detect fibre pull out in Steel Fibre Reinforced Concrete. Increased sensitivity obtained from HI could detect debonding at lower stresses. Visual observation of cracking illustrates the effect of surface roughness of fibres and asymmetric debonding (debonding on only one side of the fibres, Figure 3). Park and Jung (1988) used real time HI to study creep deformation of shale, sandstone and coal in order to validate the use of classical creep equations in these materials. They concluded that the method could be a high resolution and cheap method of monitoring time dependent behaviour.

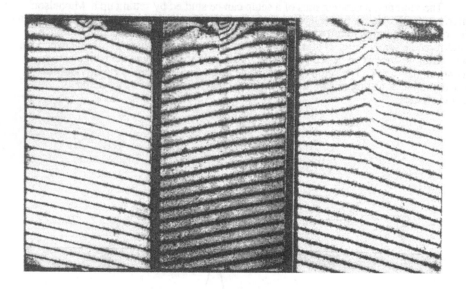

Figure 3. Fringe Patterns Showing Fibre Pullout (Stroeven et al, 1989)

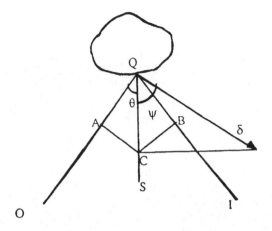

Figure 4. Sensitivity Vector for Single Observation

1.4 Quantitative measurements

The fringes in HI are caused by displacements on the object surface. Each fringe corresponds to the component of the displacement along the 'Sensitivity Vector'. The Sensitivity Vector (S) for any point on the object is the bisector of the observation direction (OQ) and the illumination direction (IQ) (Figure 4). Hence, a single observation gives only one displacement component.

Miller et. al. (1988) used this technique to find crack opening profiles of a mode I crack initiation in mortar. Since the crack opening in mode I is ideally in a single direction, it was possible to use a single observation, or a single hologram. Crack openings were found by counting fringes from one side of the crack to the adjacent point other side (Figure 5). The crack profiles were used to explain the variation in experimental observation from the predictions from a finite element analysis of a traction free crack.

In order to find the three independent displacement components of a specimen it is necessary to use three different illumination or observation directions. This could be achieved by using three separate holographic plates to observe the object surface from three different angles (Pryputniewiez, 1979).

Typically this is restricted by the cost of additional optics and the trouble involved in making three times as many holograms. It is also difficult at times to avoid rigid-body motions in-between making the three holograms for each loading stage. An alternative method developed by Maji and Shah (1990), enables measurement of displacements from a single hologram without any significant sacrifice in accuracy. This is achieved by using a single 4" x 5" (10.2 x 12.7 cm) holographic plate placed very close to the specimen surface (Figure 6). This makes each point on the object surface subtend a large angle on the holographic plate. Effectively, it is now possible to look at the object through four corners of the plate. The observed fringe pattern is different from the four corners due to the differences in the observation directions.

By counting the fringe patterns from the four corners of the same plate we have a set of four equations involving the three unknown displacement components and the four different sensitivity vectors corresponding to the four observations. A computer program

Fringe Pattern for Tensile Cracking (See Figure Below)

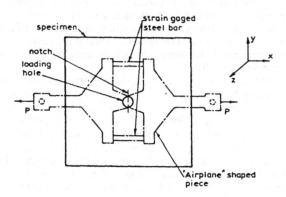

Loading Fixtures for Test Shown Above

Figure 5. Tensile Fringe Patterns (Miller et. al.)

can be used to calculate the sensitivity vectors, set up the four equations in terms of the four fringe counts and solve them by a least squares method to find the three unknown displacement components. This redundancy in observations (four observations instead of the minimum requirement of three) reduces the error if the fringe count for any single observation was wrong by ± 1/2 fringe. The in-plane displacement components (crack opening and sliding profiles) could then be evaluated. It is necessary to place the plate as close to the specimen as possible to increase the in-plane sensitivity.

One drawbacks of this technique is that, placing the plate close to the specimen makes it difficult to maintain uniform light intensities on the plate and the object . Satisfying the coherence length limitations is also more difficult.

Maji and Shah (1990) and Maji, Shah and Tasdemir (1989) used this technique to obtain crack opening and sliding profiles for a crack propagating in mixed mode. Since the crack propagation directions in mixed mode propagation are not known a priori, conventional clip gauges or crack gauges can not be used to accurately detect crack opening and sliding. That information is critical in evaluating the crack propagation criteria because of the effect of tractional forces across a crack.

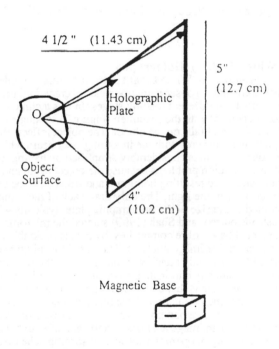

Figure 6. Four Simultaneous Observations from a Single Plate
            (Maji et al, 1990)

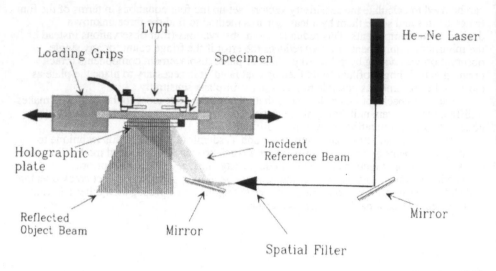

Figure 7. 'Piggyback' HI Used on Fibre Composites (Mobasher et al,1990)

1.5 Reflection (White Light) Holography

Ferrara and Morabito (1989) used 'White Light HI' to study mixed mode fracture in a four point shear test. They were able to observe microcracking, crack initiation, and formation of secondary flexural cracks at different loading stages. In this technique, the holographic plate is placed almost in contact with the specimen (Figure 7). A single expanded laser beam goes through the plate and falls on the object. The object reflects the beam which comes back to the plate and interferes with the incoming laser beam which now acts as the reference beam. The optical setup is considerably simplified by having to use a single beam. This also reduces vibration problems because the object beam and the reference beam are similarly effected. The resulting hologram and fringes can be seen without the use of laser light, i.e. in ordinary white light. The only drawback of the setup is that the viewing area is restricted to the size of the holographic plate (typically 4" x 5").

Mobasher, Castro-Montrero and Shah (1990) studied the micromechanisms of matrix fracture in portland cement based fibre composites by means of double exposure reflection (white light) HI and quantitative image analysis. They used the 'piggyback' technique where the holographic plate is rigidly attached to one point on the specimen (Figure 7). Rigid body motion is automatically eliminated, since the only relative displacement of the plate with respect to the specimen is that due to straining. A plate tilting device was developed in order to control the orientation of the interferometric fringes resulting in fringe patterns easier to analyse (i.e. fringe patterns parallel to crack patterns were avoided). The holograms were acquired into an image analysis system, and after enhancement, they were analysed for crack density, length, opening profile, and spacing. The distribution of active cracks was related to the state of damage of the specimens (Figure 8a, 8b and 8c). In Figure 8, the dotted lines represent the superposition of the cracks of the previous stages, and the solid lines represent the cracks developed during the current loading stage (active cracks). Reflection holographic interferometry was suggested as a non-destructive method of determining the state of damage in composites.

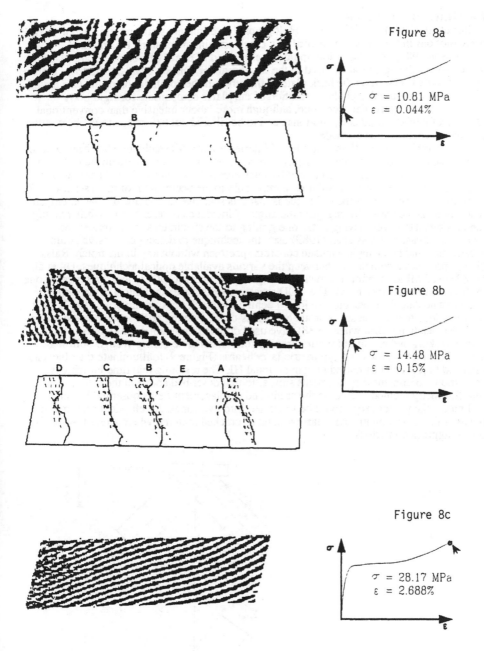

σ = 10.81 MPa
ε = 0.044%

Figure 8a

σ = 14.48 MPa
ε = 0.15%

Figure 8b

σ = 28.17 MPa
ε = 2.688%

Figure 8c

Figure 8.   Image Processing of White-Light HI to Observe
Cracking (Mobasher et al, 1990)

1.6 Other HI techniques

Miller et. al. (1988) used 'Sandwich Holography' to eliminate spurious fringes due to rigid body motions in their work on quantifying crack opening in tension. This method (Abramson 1981) uses two holographic plates instead of one. The two plates are placed one on top of the other and exposed in the usual manner. During analysis, the front plate of one exposure is placed on the back plate of the other. If the combination of the two plates (the sandwich) is tilted, rigid body rotation of the plates compensate for rigid body rotations of the object. The technique, although more labour intensive than conventional HI, is useful where rigid body motions need to be filtered out before accurate observations and fringe counting can be made.

A combination of HI and the Moire´ Interferometry (MI) technique has been used by many researchers for quantitative measurements of strain and displacement field. In this technique, the moire´ grating is created by interference between two laser beams which causes the grating to have a resolution comparable to the wavelength of the laser used. The consequent advantages of the technique is that it is sensitive to in-plane displacement, and resolution can be varied by changing the angle of incidence of laser beams while making the gratings. However, having to fix one grating to the specimen surface makes the technique invasive. Raiss et al. (1989) used the technique to detect progressive strain localization and cracking in a tensile concrete specimen without any initial notch. Raiss (1986) provides a comparison between the various available optical techniques. Du et al. (1987) used MI to monitor the crack opening displacement (COD) in and the process zone in three point bend specimens. The COD measurements were input into a finite element code to calculate the crack closure forces behind the crack tip. Cedolin et al. (1987) used the technique to evaluate stress-strain and stress-crack opening relations in tension.

Regnault and Bruhwiler (1988) have used double illumination HI to detect microcracking and extent of fracture process zone in concrete during Wedge Splitting Tests. This method uses two symmetric laser beams (Figure 9) to illuminate the object as opposed to just one beam used in conventional HI. The resulting real time HI fringes correspond to in-plane displacements similar to that described above in the moire´ interferometry method. Based on the strain and displacement fields observed by this technique, the process zone could be separated into two zones. An elliptical region which behaves like a continuum and a narrow band of cracked material where stress transfer is due to aggregate interlock.

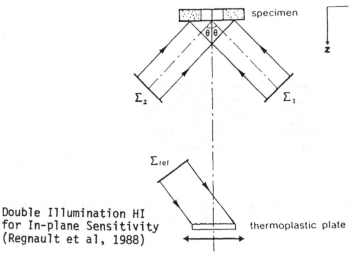

Figure 9.   Double Illumination HI
             for In-plane Sensitivity
             (Regnault et al, 1988)

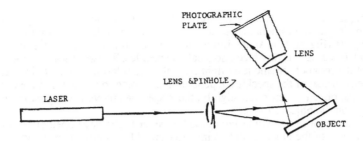

Figure 10. Set-up for Speckle Photography

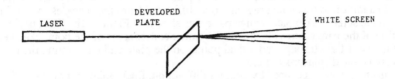

Figure 11a. Set-up for Observing Speckle Fringes

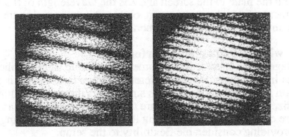

Figure 11b. Typical Fringe Patterns Observed

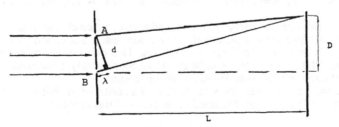

Figure 11c. Bragg's Diffraction

2 Speckle Photography (SP)

In this technique, the specimen or object surface is illuminated with a laser light and a setup similar to a photographic camera is used to focus the image of the object on to a holographic plate (Figure 10). When illuminated by coherent laser light, the object surface seems to contain numerous tiny black dots called speckles. These dots are caused by diffraction of the monochromatic laser light from the irregularities on the object surface. The characteristic size of the speckles depend on the surface roughness and the optical setup. When the object is deformed slightly, these speckles move corresponding to displacements on the object surface.

The plate is exposed at two different loading stages. The developed plate thus contains the image of the object with two invisible speckle patterns corresponding to the two loading stages. When an object point moved, the dots or speckles moved correspondingly. These two images of the object therefore contains two speckle patterns that are displaced slightly from each other due to the displacement of each point.

To read the displacements at individual points on the object, the set up in Figure 11a is used. An unexpanded (narrow beam from the laser source) laser light is passed through the point. A white screen is placed behind the plate. The direct laser ray falls on the screen. In addition to this direct ray, the light diffracted from the speckles form a fringe pattern on the screen. A typical fringe pattern is shown in Figure 11b. The spacing and orientation of the fringes provide the displacement components of the point. Hence, by passing the laser light through individual points on the plate and interpreting the fringes one can determine the displacement field.

This phenomena is caused by Bragg's diffraction. Each point that the laser light passes through, contains numerous speckles each of which are displaced by the same amount x (Figure 11c). These speckles diffract the incident laser beam causing an interference pattern on the screen behind. Fringe spacing D is obtained as $D = \lambda L / x$ where L is the distance from the plate to the screen and λ is the wavelength of the He-Ne laser. The actual displacement on the object points are obtained by accounting for the magnification in the image obtained by the camera.

2.1 Comparison with holographic interferometry

The obvious advantage of the speckle interferometry technique is its simplicity and consequent financial implications. Since interference of two laser beams is not involved, the vibration isolation requirements are far less stringent. Vibration isolations limits are determined by the size of the speckles which are typically much larger than the laser's wavelength. The size can be varied by changing the aperture of the imaging lens, object to lens distance etc., providing considerable flexibility to the setup.

It is sensitive to in-plane displacements and quite insensitive to out-of-plane motions, a consequence of the shape of the speckles. Sensitivity is determined by two major factors.

a) The speckle size should be much smaller than the displacement of corresponding points. Otherwise the points A and B on Figure 10c would overlap and no diffraction pattern will occur.

b) Speckle size can be altered as mentioned above, hence changing sensitivity.

It is also possible to use a plate with much lower spatial resolution (lines /mm) than those used for HI because of the larger speckle size. This means that a more sensitive plate can be used and the exposure time can be cut down considerably. Low exposure time also means less vibration problems and isolation requirements.

Despite these advantages, the SP technique is an order of magnitude less sensitive than the HI or the SI technique discussed in section 3 of this chapter.

2.2 Applications to study concrete microfracture

De-Backer (1975) used SP to measure displacement fields in notched PMMA beams and reinforced concrete beams in an effort to study the behaviour of concrete. Extensive study of crack opening and displacements in the process zone under tensile fracture have been done by Ansari using SP (1984, 1987). Accurate description of the fracture process zone was possible. A small scale fracture zone due to debonding of aggregates, a few mm. wide, was found to exist prior to unstable growth. The instrumentation included an automated system for reading, recording and analysing the fringe patterns. A small servo controlled machine was mounted on the vibration isolation table. Maji and Shah (1988) used the technique to identify the mode of cracking at the aggregate interface under compression. Their study concluded that although the crack initiation is in mixed mode condition, tensile cracking soon becomes the dominant mode.

3 Speckle Interferometry (SI)

This technique is based on the speckle photography described in the previous section. However, unlike the single beam of light used to illuminate the object in SP, two beams are used (Figure 12). These two beams are symmetric i.e. they have the same incident angle on the specimen. The speckle photograph hence obtained is caused by the interference of the two beams on the object surface. If the specimen moves in its own plane, the path length for one beam increases while that for the other decreases. The speckle pattern would change accordingly. For those points on the specimen for which the path length changes by an integral multiple of the laser wavelength, the speckle pattern does not change. For those points where the path length changes by an odd multiple of $\lambda/2$, the speckle pattern reverses itself, i.e the dark spots on the speckle pattern become bright and vice versa. The resulting application is similar to HI except that speckle photographs are made at different stages of loading. Both real-time and double-exposure applications are possible and in-plane displacements are visible on the speckle photographs as fringes of high and low contrast. Out-of-plane displacements cause the same path length change for the two incident beams and therefore cause the same speckle pattern. That makes the technique insensitive to out-of-plane motions.

3.1 Comparison with the other techniques

The optical setup is similar to that used in HI except for the imaging system replacing the holographic plate. Holographic plates are typically used in SI however, such high resolution films are not necessary. Unlike in SP where the speckle size had to be much smaller than the displacement for diffraction to occur, in SI, the speckle size has to be much larger than the displacement. This could mean either that smaller displacements can be detected or that the speckle size could be larger. SI has the same sensitivity as HI and has the advantage of being sensitive to in-plane displacement as opposed to the usual out-of-plane sensitivity of HI. That makes it about an order of magnitude more sensitive than SP. Also a single observation is enough to obtain in-plane displacement data unlike at least three observations needed for HI (section 1.4 of this chapter). Vibration isolation requirements are as stringent as those in HI. However, as the speckle size could be larger, vibration isolation requirements on the imaging system are less since using less sensitive film (lines / mm) it is possible drastically to cut down on the exposure time.

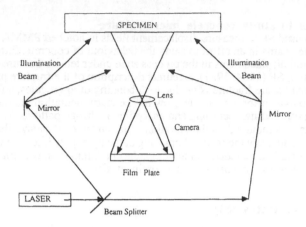

Figure 12. Simplified Set-up for Speckle Interferometry

3.2 Electronic speckle pattern interferometry (ESPI)

In the SI technique described above, a speckle photograph is made at some initial loading stage and any changes in speckle pattern is observed as a comparison to that initial photograph. If the imaging plate is replaced by a video camera, the speckled images can then be recorded on a video. Successive changes of speckle patterns would be continually monitored. The information can be stored on an image processing system which could compare any two speckle patterns at a later point of time and produce the same fringe patterns as would otherwise be observed by SI. Replacing the holographic plate with a video camera is possible because of the large size of speckles involved which can be captured in sufficient details with the resolution of normal video cameras. Various engineering and nondestructive testing applications have been described by Jones and Wykes (1983).

Although the instrumentation is more expensive, it removes the process of plate developing which is time and labour intensive, and also critical for getting good quality images. Since the data is acquired at TV speed (typically 30 frames per second, the vibration isolation requirements are far less stringent than for conventional HI or SI techniques which need long plate exposure times. This makes ESPI more usable in an industrial environment, and makes it a very versatile technique. All data are automatically digitized and could be analysed at any later point of time. In SI, the initial loading stage is fixed, once the first exposure is completed. In ESPI, any two stages of loading can be compared by numerical manipulation of the digitized data for those two loading stages. The technique is also more suited for remote sensing purposes.

Maji and Wang (1990) have used ESPI on Compact Tension specimens of Indiana Limestone. They were able to measure pre-peak crack propagation, measure opening profiles along the propagating crack and the preexisting notch. This quantitative data was then used to generate information on the stress transfer due to ligament connections across the propagating crack.

Hansen (1989) used a similar technique called TV Holography to study flexural cracking of notched concrete beams. This method uses a single object beam and a reference beam which is directly acquired by the TV camera. Cracking was observed in real time with an accuracy of 0.1 micron. Cracking was found to initiate at about 80 % of ultimate load. Aggregate particles were observed to arrest crack growth and promote crack branching. Quantitative measurement of cracking was however not done in that research.

3.3 Shearography

The SI and HI technique provides only the displacement data. To get an idea of the stress field across a crack, stress concentrations at the crack tip and other physical phenomena, very often, the necessary data item is strain. Obtaining strains from a displacement field by a point by point differentiation is a laborious process and often leads to large errors. Inaccuracies in displacement data magnifies during differentiation. The need for a direct strain measuring technique has led to another variation of the SI technique. In the most simple case, the conventional SI setup is used. A prism is used to cover half the imaging lens. This, in effect, causes the images from two different object points to be focused on the same point on the imaging plate (Figure 13). The speckle pattern at point A is hence caused by the interference of the light from points B and C. Fringe pattern on the image plane now corresponds to the relative displacements of points (such as B and C) on the object plane. Differentiation from displacement field to strain field is hence carried out optically and the resulting image provides full-field strain data with a sensitivity that can be varied widely by changing the imaging system and the prism.

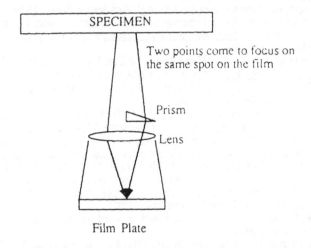

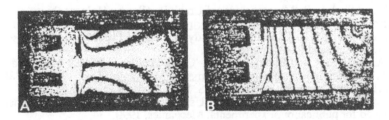

Figure 13.   Shearography Set-up and Fringe Pattern (Erf, 1974)

References

Abramson N. 'The Making and Evaluation of Holograms', Academic Press, 1981

Ansari F.'Experimental Analysis of Fracture in High Strength Cementitious Composites', in 'Very High Strength Cement Based Materials', MRS publication V 42, 1984.

Ansari F. 'Microcracked Concrete in Direct Tension', ACI Materials J., V 84, No. 6, Nov-Dec, 1987, pp. 481-490.

Cedolin L., Dei Poli S. and Iori I. 'Tensile Behavior of Concrete', ASCE J. of EMD, V113 (3) , 1987, pp. 431-449.

DeBacker L. C. 'In-plane Displacement Measurement by Speckle Interferometry', Nondestructive Testing, August 1975, pp.177-180.

Du J. J., Kobayashi A. S. and Hawkins N. M. 'Fracture Process Zone of a Concrete Fracture Specimen', Proc. SEM/RILEM Int. Conf. on Fracture of Concrete and Rock, Houston, TX, June 1987, Edited by S. P.Shah and S. E. Swartz.

Ebbeni J., editor, 'Industrial Applications of Holographic Nondestructive Testing', SPIE Proc. V 349, Brussels, May 1982.

Erf R. K. 'Holographic Nondestructive Testing', Academic Press, 1974.

Ferrara G. and Morabito P. 'A Contribution of the Holographic Interferometry to Studies on Concrete Fracture', Proc. of Int. Conf. Fracture of Concrete and Rocks, Cardiff, UK, Sept. 1989.

Hansen E. A. 'A Holographic Real Time Study of Crack Propagation in Concrete', Cement and Concrete Research, V 19, 1989, pp. 611-620.

Hariharan P. 'Optical Holography, Principles, Techniques and Applications', Cambridge University Press, 1984.

Iori I. 'Process Zone in Normal Concrete', RILEM 90-FMA Report 'Fracture Mechanics of Concrete, Applications', Chapter 3 'Material Properties', 1988.

Jacquot P. and Rastogi P. K. 'Speckle Metrology and Holographic Interferometry Applied to the Study of Cracks in Concrete', in Fracture Mechanics of Concrete, Edited by F. H. Wittman, Elsevier Publishers B. V. , Amsterdam, 1983.

Jacquot P. 'Interferometry in Scattered Coherent Light Applied to the Analysis of Cracking in Concrete', Fracture Mechanics of Concrete, edited by A. Carpinteri and A, Ingraffea, Martinus Nijhoff Publishers, 1984.

Jones R. and Wykes C. 'Holographic and Speckle Interferometry', Cambridge University Press, 1983.

Maji A. K. and Shah S. P. 'Application of Acoustic Emission and Laser Interferometry to Study Microfracture in Concrete', ACI-SP 112 Nondestructive Testing, 1988, pp. 83-109.

Maji A. K., Shah S. P. and Tasdemir M. A. 'A Study of Mixed Mode Crack Propagation in Mortar Using Holographic Interferometry', Proc. SEM Spring Conf., Boston, MA, May 1989.

Maji A. K. and Shah S. P. 'Measurements of Crack Profiles by Holographic Interferometry', to be published in Experimental Mechanics, 1990, paper no. 3872.

Maji A. K. and Wang J. L. 'Experimental Study of Fracture Processes in Rock', submitted to Rock Mech. and Rock Engng., 1990.

Miller R. A., Shah S. P. and Bjelkhagen H. I. 'Crack Profiles in Mortar Measured by Holographic Interferometry', Experimental Mechanics, Dec. 1988, pp. 388-394.

Mobasher B., Castro-Montrero A. and Shah S. P. 'A Study of Fracture in Fibre Reinforced Cement-Based Composites Using Laser Holographic Interferometry', Accepted for publication, Experimental Mechanics, 1990.

Park D. and Jung J. 'Application of Holographic Interferometry to the Study of Time-Dependent Behavior of Rock and Coal', Rock Mechanics and Rock Engineering. 21, 1988, pp. 259-270.

Pryputniewiez R. 'Holographic Determination of Rigid Body Motions and Application of the Method to Orthodontics', Applied Optics, V 18, p. 1492, 1979.

Raiss M.E., Dougill J.W. and Newman J. B. 'Observation of the Development of Fracture Process Zone in Concrete', Proc. Int. Conf. on Fracture of Concrete and Rock, Cardiff, UK, Sept. 1989. pp. 243-253.

Raiss M.E. 'Observation of the Development of of Fracture Process Zones in Concrete Under Tension', PhD Thesis, University of London, 1986.

Regnault P. and Bruhwiler E. 'Holographic Interferometry for the Determination of Fracture Process Zone in Concrete', Proc. Int. Conf. on Fracture and Damage of Concrete and Rock, Vienna, July 1988.

Smith H.M. 'Principles of Holography', Wiley Interscience, 1969.

Stroeven P., Burakiewicz A. and Bien J. 'Holographic Interferometry Approach to a Study of Debonding in Plain and Steel Fibre Reinforced Concrete', Proc. SEM Spring Conf., Boston, MA, May - June 1989, pp. 689-696.

Tasdemir M. A., Maji A. K. and Shah S. P. 'Crack Initiation and Propagation in Concrete Under Compression', to appear ASCE J. of EMD, May 1990.

Tasdemir M. A., Maji A. K. and Shah S.P. 'Mixed Mode Crack Propagation in Concrete Under Compressive Loading', Proc. Int. Conf. on Fracture of Concrete and Rock, Cardiff, UK, Sept. 1989.

Vest C. M. 'Holographic Interferometry', Wiley , 1979.

SUBJECT INDEX

Accelerometers 207
Acetylcellulose film 234, 237
Acoustic emission analysis 107, 209,
    241–5, 254
Aggregate size 22, 101, 118, 159, 167,
    177, 216, 240
    relation with fracture
        toughness 118–20
ANSYS 165
Arcan specimen 175–7, 178
Axial compression 165, 167

Backscattered electron imaging 237
Beam Splitters 263
Beams, local damage 204–5
Bending tests 199–200
Biaxial test rigs 173–4
Blunt crack model 89
Bragg's diffraction 273, 274
Brazillian Splitting Test 160
Brittleness number 137–8, 162, 163

Charpy impact test 199, 200, 201
Circumferentially notched compact
    beam 167
Cohesive crack models 170
    compared 37–47
    determination of parameters 47–52
    with linear softening 43–5
    with quasi-exponential softening
        42–3
Compliance measurements 32–3, 95,
    108–21, 123, 125, 233–4
Compression tests, model
    concrete 182, 185
Compressive load, fracture 179–86
Compressive preloading 216
Compressive shear 175–7
Compressive strength 26
Compressive stress 185–6

Concrete, large structures; *see* Large
    structures
Crack detection, holographic
    interferometry 266
Crack gauges 221
Crack initiation 153, 167, 266, 275
    rate 214–15
Crack length, effective 109–10, 114,
    115, 116, 117, 122
Crack mouth opening displacement
    (CMOD) 2, 3, 5, 7, 22, 47, 48,
    124, 150
Crack opening displacement 92, 96,
    98, 99, 102, 108, 244, 267, 272,
    275
    traction free 217
Crack propagation 136, 137, 139, 142,
    162, 165, 211, 213, 241, 266, 269,
    276
    velocity 215, 220–1
Crack tip opening displacement 3, 8,
    31, 47, 48, 49, 89, 123, 124, 126
    150
CRACKER 165, 167
Cracking gauges 205, 207
Cracks
    in compressive stress fields 185–6
    penny-shaped 211–13
    size transitions 136–8
Critical crack length, effective crack
    model 23–5
Critical stress intensity 2–3, 6, 8, 54,
    89, 147
    effective crack model 22–30
Critical stress intensity factor; *see*
    Fracture toughness
CT-specimen 211

Deconvolution techniques 242
Deformation measurement 210

280

AUTHOR INDEX

Milton Keynes UK
Ingram Content Group UK Ltd.
UKHW040447071024
449327UK00020B/1049